Dietrich Klemm, Rosemarie Klemm

THE STONES OF THE PYRAMIDS

German Archaeological Institute
Cairo Department

Dietrich Klemm, Rosemarie Klemm

THE STONES OF THE PYRAMIDS

Provenance of the Building Stones of the Old Kingdom Pyramids of Egypt

DE GRUYTER

ISBN 978-3-11-022123-7

Library of Congress Cataloging-in-Publication Data

Klemm, Dietrich D., 1933-
The stones of the pyramids : provenance of the building stones of the Old Kingdom pyramids of Egypt / by Dietrich Klemm and Rosemarie Klemm.
p. cm.
Includes bibliographical references and index.
ISBN 978-3-11-022123-7 (hardcover : alk. paper)
1. Pyramids–Egypt. 2. Pyramids–Egypt–Design and construction. 3. Building stones–Egypt–History. 4. Building stones–Egypt–Analysis. 5. Limestone–Egypt–Analysis. 6. Petrology. 7. Geochemistry 8. Quarries and quarrying–Egypt–History. 9. Egypt–Antiquities. 10. Egypt–History–To 332 B.C. I. Klemm, Rosemarie. II. Title.
DT63.K58 2010
932'.012–dc22
2010002750

Bibliographic information published by the Deutsche Nationalbibliothek
The Deutsche Nationalbibliothek lists this publication in the Deutsche Nationalbibliografie; detailed bibliographic data are available in the Internet at http://dnb.d-nb.de.

Cover image: R. Klemm/D. Klemm „Gizeh plateau, pyramid of Khafre with pyramid temple“
Printing: NEUNPLUS1 GmbH, Berlin
♾ Printed on acid-free paper
Printed in Germany
www.degruyter.com

Content

Preface

The aim of this investigation is focused on the provenance determination of both core and casing material of the Old Kingdom stone pyramids and their temple complexes. For this purpose, rock samples in sufficient number were taken from the pyramids' construction blocks to achieve a statistical distribution pattern of their various quarry sites. Sampling was always carried out in the presence of and in agreement with an authorised representative of the Egyptian Supreme Council of Antiquities (SCA). These pyramid rock specimens were compared with rock samples from known or assumed quarry sites in the surrounding area of the pyramids. Or, concerning the casing and backing stones[1] from the reign of Snofru on, from the extended quarry sites that can be found in the eastern escarpment of the Nile valley and were well known at Mokattam, Tura, and Maasara (Wadi Hof).

The pyramids under discussion are situated along the Nile valley at the edge of the Western Desert plateau between Abu Roash in the north and Meidum in the south (Fig. 1). In the present study, the following pyramid complexes (listed from north to south) will be referred to in detail:

Abu Roash:	Djedefre : 4th dynasty.
Gizeh plateau:	Khufu, Khafre, Menkaure: 4th dynasty.
Zawiyet el-Aryan:	Khaba: 3rd dynasty (?).
	Baka (or Nebka): 4th dynasty (?).
Abusir:	Sahure, Neferirkare, Neferefre, Khentkaus II, Niuserre: 5th dynasty.
Saqqara-North:	Djoser, Sekhemkhet: 3rd dynasty.
	Userkaf, Menkauhor (?), Unas: 5th dynasty.
	Teti: 6th dynasty.
Saqqara-South:	Shepseskaf (Mastabat el-Faraoun): 4th dynasty.
	Djedkare Isesi: 5th dynasty.
	Pepi I, Merenre, Pepi II: 6th dynasty.
Dahshur:	Snofru: Bent Pyramid and Red Pyramid: 4th dynasty.
Meidum:	Huni: 3rd dynasty (?), Snofru: 4th dynasty.

Apart from this list, we shall deal with the various pyramids in chronological order. This gives a better view on developments in the usage of building materials (Tab. 1). However, it has the disadvantage that, particularly in the Saqqara area, repetitions are unavoidable within the geological-petrographic descriptions.

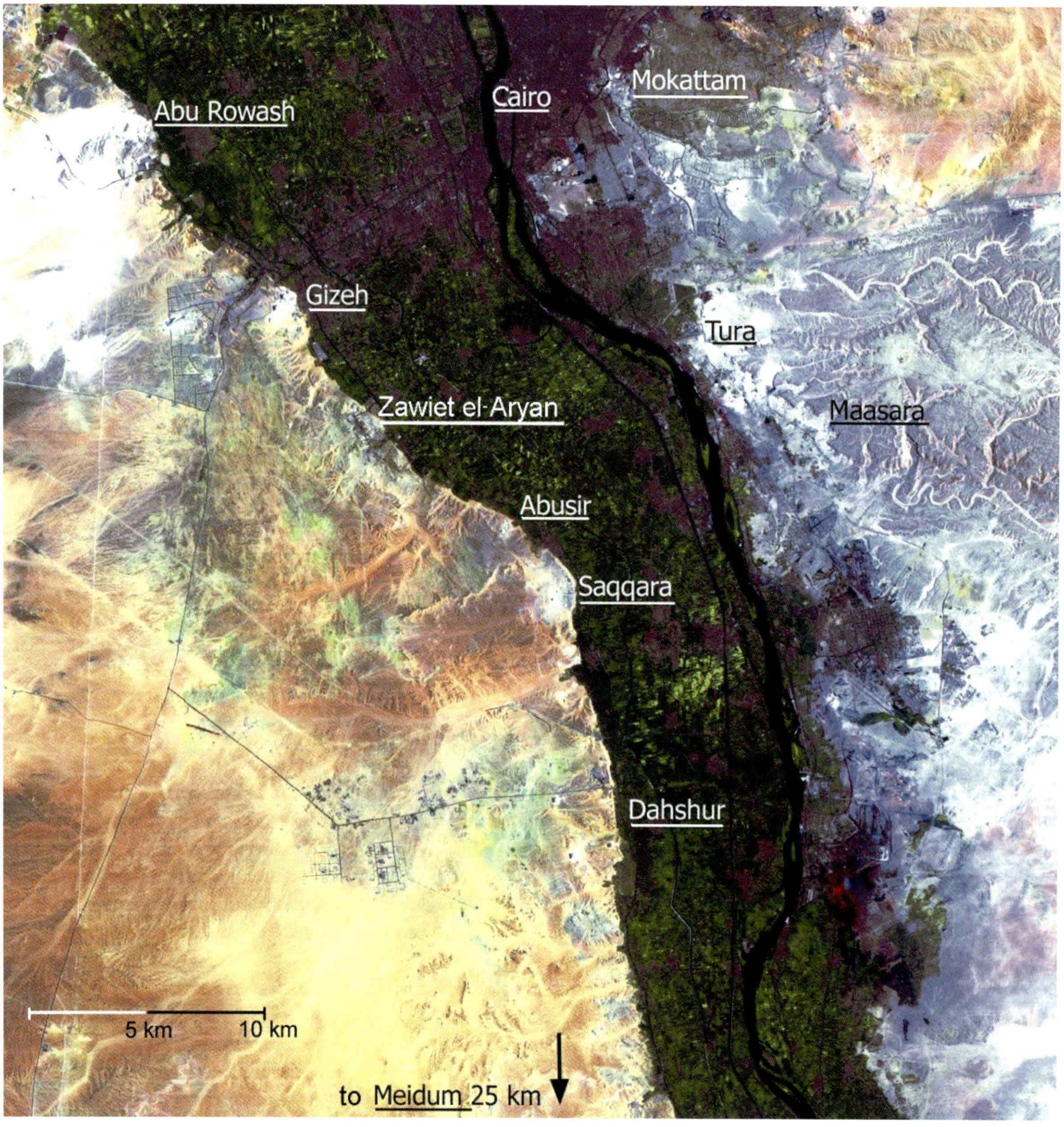

Fig. 1: Satellite image (TM scene 176/39) of the Old Kingdom pyramid districts from Abu Roash until south of Dahshur and the quarry areas Mokattam, Tura, and Maasara. Note that the vicinity of the pyramids as well as the quarry areas appear whitish as a result of working operations.

Acknowledgements

Firstly, we would like to express our deepest gratitude to the German Volkswagen-Stiftung, which made our investigations possible with generous financial support and which accompanied the entire program with most helpful goodwill. Here, special thanks should be given to MARIE LUISE ZARNITZ and GÜNTER DEGE for their constant encouragement and assistance.

During our fieldwork in Egypt, we had the great pleasure of the close cooperation and hospitality of the German Archaeological Institute in Cairo. We are especially indebted to WERNER KAISER and RAINER STADELMANN. The late WINFRIED BARTA kindly allowed access to the facilities of the Institute of Egyptology, University of Munich, and helped whenever needed. The Institute of General and Applied Geology of the University of Munich made laboratories and technical assistance available, which is gratefully acknowledged. Further, we are very much indebted to the Egyptian Supreme Council of Antiquities (SCA) and here especially to General Director MAHMUD ABDEL RAZIQ and to ZAHI HAWASS, at the time of our fieldwork General Director of the Gizeh and Saqqara area, and to the inspectors MANSUR B. RADWAN, MAGDI HASSANAIN ALI, and HUSSEIN LASCHIN of Saqqara. For analytical assistance, we are grateful to ALFRED UTKE (Department of Geology, Technical University of Berlin), for sample preparation MARIANNE WERNER and EDITH WEIHRICH, and for intensive office help HEIDE SCHÄPE. Finally, we would also like to thank CATHERINE JONES for improving the English text version.

Introduction

During the Old Kingdom (~ 2700–2150 BC) the most striking monuments of Egypt were constructed, namely the pyramids and their adjacent temple complexes.

Egyptian pyramids are to be regarded as royal tombs, ideally completely constructed during the lifetime of the pharaoh, to accommodate the royal corpse in a stone sarcophagus, which had already been placed within the burial chamber during the pyramid's construction. For the organization of the funeral and subsequent rituals, several further building complexes were established: A valley temple at the foot of the desert escarpment acted as the ceremonial entrance into the pyramid complex. In general, this valley temple is connected to a causeway for the processional transport of the royal body towards the funeral (mortuary) temple next to the pyramid, and finally into the pyramid itself. Occasionally, a way station for ceremonial purposes occurs at the side of the causeway. As a rule, the pyramid and the mortuary temple are framed by an enclosure wall. Quite a number of the pyramids contain distinctly smaller satellite pyramids, reserved as both tombs for the queens or for a cult not yet clearly understood.

Within the further surrounding area of these pyramid complexes, usually the tombs of royal family members as well as of higher officials were constructed, so that altogether building complexes of enormously high expenditure in terms of material and supplies gradually developed.

As the pyramids are the most fascinating monuments of Ancient Egypt, almost uncountable detailed publications are existent all over the world. There is no need to repeat them again. In the following chapters, we will refer only to this somewhat generalized picture of a pyramid complex as described above, and the archaeological descriptions will not be extended into detail.

For the execution of these enormous constructions, the supply of large rock quantities and the coordinated procurement from suitable quarries was the most important requirement. Both the technical procedures and the origin of the rocks used for building the pyramids remain the subject of various considerations and computations, which flow sometimes into absurd speculations of many non-professionals, the so-called "pyramidologists".

As authentic sources are lacking from the working sphere of the pyramid architects, no generally accepted model for the pyramid construction exists as yet. However, the question about the provenance of the building stones can now be answered with the help of scientific comparative investigations of quarry samples in relation to samples of the building stones, particularly since in the past years the assumed ancient Egyptian quarry sites were examined petrographically and geochemically by the present authors.

The discovery of the material "stone" and its use for monumental constructions was vigorously developed at the beginning of the 3rd dynasty. One might assume that the consolidation of the ancient Egyptian state was conducive to this technical revolution with the understanding of Egypt as a "state out of stone"[2].

Analytical procedure

The comparison of the pyramid and quarry stone material was made by applying petrographical and geochemical methods.

The petrographical methods necessitate a short description of the rock feature firstly in macroscopical terms, secondly under a 10 times magnifying hand lens, and finally under a polarizing petrographic microscope to define the mineralogy, structure, and fossil content of the specimens. Hereby, a sample slab of about 5–6 cm^2 was mounted with araldite onto a glass plate and grinded down to a thin section of 30–25 µm thickness. Thus, the sample becomes transluminent and the content can be studied with a high-resolution polarizing microscope. This allows a detailed study of the rock fabric including structure (micro-facies), mineral identification, and the recognition of fossils or fossil remains even in nano-sizes. The application of higher nano-methods such as scanning electron microscopy (SEM), as demonstrated by KLEMM/KLEMM[3] for provenance attributions of extremely small specimens from archaeologically highly valued objects, was in the case of this investigation unnecessary, as sufficient material for effective results was always available.

Parallel to the petrographical comparisons, geochemical analyses were undertaken using the simple and low-cost method of induced coupled plasma spectroscopy (ICP), available in almost all geochemical laboratories worldwide.

As already mentioned, all pyramids of the Old Kingdom were constructed generally using two basically different rock types, one for the core material, composing more than 90 % of the volume, and the casing and backing masonry. Both types could consist of material from quite different quarry sources. The core material, however, was extracted as often as possible from sources close to the construction site. The casing and backing material, with the exclusion of the 3rd dynasty pyramids, came from the quarries at Tura and Maasara, and in case of Khufu's pyramid, partly also from Mokattam on the opposite bank of the river Nile.

We procured a sufficient amount of samples from both the core and casing material of each pyramid. We repeated this in the rock outcrops that come into consideration as ancient quarry sites. The results of chemical analysis carried out on samples from both the pyramid and possible quarry sites were compared and, based on similarities of content, the provenance of the pyramid blocks could be determined.

The limestone used for the construction of the Old Kingdom pyramids is composed of mineral calcite (calcium carbonate with the ideal chemical formula $CaCO_3$). However, almost all Egyptian limestones contain, in more or less amounts, clay minerals. Particularly the limestones south of the Gizeh plateau contain different fractions of sand components, mostly of quartz, but also feldspar and other minerals.

Thus, as generally all limestones are composed mainly of calcium carbonate, the analysis of calcium gives little indication of a chemical differentiation. Consequently, this element was excluded from analytical investigations.

By means of several geochemical pre-investigations, it could be established that the trace elements magnesium (Mg), iron (Fe), manganese (Mn), strontium (Sr), barium (Ba), and subordinated zinc (Zn) give an optimum of differentiation possibilities. We are aware of the fact that we were using rather mobile elements, but comparative tests of archaeologically unambiguous, well-defined quarry areas, such as the Menkaure quarry next to his pyramid, with the core material of the monument itself yielded satisfying results, thus giving us a secure base for the analytical procedure applied.

A first attempt to analyse at least parts of our sample bank, which is now stored at the British Museum, was undertaken by the late LORE HOLMES from the Brookhaven National Laboratory. She used the Neutron Activation Analysis (NAA) method including the trace elements Na, Rb, Sr, Ba, Cr, Fe, Co, Sc, La, Ce, and Eu of various limestones to successfully correlate Coptic sculptures with their provenance quarries. Unfortunately, these very interesting results are not yet published, due to LORE's inconsolable early death.

Caustic critics may regard our method as unsatisfactory, inasmuch as the clay and sand fractions were not treated further or chemically analysed. It should be emphasized, however, that the goal of this investigation was focused pragmatically on the assignment of the sampled building material to certain quarry provenances and not on completeness of chemical analysis. The more, as we were bound to a strictly limited budget.

For analytical treatment, approx. 100 g of the samples were reduced to pea size and subsequently milled for one minute in a WIDIA-disc mill. From the rock powder, 5 g were dissolved with 2-normal hydrochloric acid, filtered, and analysed. The filter remains were dried and separately weighed.

The actual analytical treatment followed the ICP-method (Induced-Coupled-Plasma), used at the Department of Geology, Technical University of Berlin. This analytical procedure guarantees very low limits of detection, in the case of the above mentioned elements within a low ppm-order (ppm = parts per million), which is of great importance for the attribution of a certain sample of building material to a well-defined quarry site. Furthermore, the ICP-method has reached such an advanced stage of development that, independent of the instrumental type, its results are internationally compatible. This makes us hopeful that the data presented here may be regarded as being reproducible. In any case, a sufficient amount from almost all samples has been retained for further, more sophisticated analytical methods.

Whilst climbing the pyramids listed above, selected rock samples were taken, considering that in relation to the mass of the building constructions a statistically representative number of samples was necessary. From the Khufu pyramid for instance, 130 samples were taken from the north-eastern edge: 80 at the lower third, 40 at the middle third, and 10 at the uppermost. At the south-western edge, additional 50 samples were taken from the lower part including samples from the western and southern basal zone. Apart from a wide spectrum of more precious stone types used in the internal chambers of the pyramids and their temples such as red granite, granodiorite, tonalite, calcite-alabaster, quartzite, basalt, and greywacke, the main mass of the core and casing masonry in general is limestone. However, these limestones show a rich variation of compositions, whereby a conspicuous bimodal division between the rock qualities of the core masonry and the outside casing (including "backing stones") is recognizable in all pyramids. Apart from this general bimodality, blocks of rock material belonging to the casing type could be found, in some cases, within the core masonry, but never vice versa.

During sampling, a compromise between the aspects of monument conservation and a minimum sample size, suitable for scientific research, was an utmost priority. It was necessary to procure a thin petrographic section from most samples and a sufficient amount of powder from each sample for geochemical analysis, statistically representing its average composition. In agreement with the Egyptian Supreme Council of Antiquities, samples of a size corresponding to a cube of about 4 cm were taken.

For each pyramid complex, a letter code was selected to precede the sample number, whereas the quarry samples are marked QS + number. The sample points were marked as exactly as possible on photos, published layout plans, or position sketches and listed in a continuous sample protocol.

Table 2 gives the amount of samples taken from each pyramid complex, divided into limestone for the core and casing masonry and other special rock types. Further attention was also paid to collect random mortar samples; the results of their analysis have been published in part[4]. Especially within the temple areas, the used rock types were mapped in ground plans which already exist in Egyptological literature. Due to space restrictions, these plans could not be published here, but they will be available together with the samples bank at the British Museum Department for Scientific Research in London.

Tab. 1: List of rock types used within the various Old Kingdom pyramid complexes (in chronological order)

Djoser	Loc. limest.	Fine limest.	Calc.alab.	Red granite	Granodiorite	Quarzite	Greywacke	Basalt
Vall. temple								
Causeway	X							
Pyr.temple			X					X
Pyramid	X	X						
Substructure				X				
Sekhemkhet	Loc .limest	Fine limest.	Calc.alab.	Red granite	Granodiorite	Quarzite	Greywace	Basalt
Vall. temple								
Causeway								
Pyr.temple								
Pyramid	X	X						
Substructure			X					
Khaba	Loc. limest	Fine limest.	Calc.alab.	Red granite	Granodiorite	Quarzite	Greywacke	Basalt
Vall. temple								
Causeway								
Pyr.temple								
Mortuary temple								
Pyramid	X							
Substructure								
Nebka (?)	Loc. limest.	Fine limest.	Calc.alab.	Red granite	Granodiorite	Quarzite	Greywacke	Basalt
Vall. temple								
Causeway								
Pyr.temple								
Pyramid	X							
Substructure				X				
Huni (?) Meidum	Loc. limest.	Fine limest.	Calc.alab.	Red granite	Granodiorite	Quarzite	Greywacke	Basalt
Vall. temple								
Causeway								
Pyramid	X	X						
Pyr.temple								
Substructure								
Snofru Meidum	Loc .limest	Fine limest.	Calc.alab.	Red granite	Granodiorite	Quarzite	Greywacke	Basalt
Vall. temple		X						
Causeway								
Pyr.temple								

Pyramid	X	X						
Substructure								
Snofru Bent	Loc. limest.	Fine limest.	Calc.alab.	Red granite	Granodiorite	Quarzite	Greywacke	Basalt
Vall. temple	X	X		X				
Causeway								
Pyr.temple								
Pyramid	X	X						
Substrucure				X				
Snofru Red	Loc. limest.	Fine limest.	Calc.alab.	Red granite	Granodiorite	Quarzite	Greywacke	Basalt
Vall. temple	X	X						
Causeway	X							
Mortuary temple	X	X						
Pyramid	X	X						
Substructure				X				
Khufu	Loc. limest.	Fine limest.	Calc.alab.	Red granite	Granodiorite	Quarzite	Greywacke	Basalt
Vall. temple	X	X						
Causeway	X							
Pyr.temple		X	X					X
Pyramid	X	X						
Substructure				X				
Radjedef	Loc. limest.	Fine limest.	Calc.alab.	Red granite	Granodiorite	Quarzite	Greywacke	Basalt
Vall. temple								
Mortuary temple	X	X	X	X	X	X		
Pyramid	X			X				
Substructure	X	X		X				
Pyramid				X				
Substructure				X				
Khafre	Loc. limest.	Finelimest.	Calc.alb	Red granit	Granodiorite	Quarzite	Greywack	Basalt
Valley temple	X		X	X	X			X
Causeway	X							
Mortuary temple		X	X	X	X			
Pyramid								
Burial chamb				X				
Menkaure	Loc. limest.	Fine limest.	Calc.alab.	Red granite	Granodiorite	Quarzite	Greywacke	Basalt
Vall. temple	X							

Causeway		X						
Pyr.temple	X			X	X			
Pyramid	X	X		X				
Substructure				X	X			
Shepseskaf	Loc. limest.	Fine limest.	Calc.alab.	Red granite	Granodiorite	Quarzite	Greywacke	Basalt
Vall. temple								
Causeway								
Pyr.temple	X							
Pyramid	X			X				
Substructure				X			X	
Userkaf	Loc. limest.	Fine limest.	Calc.alab.	Red granite	Granodiorite	Quarzite	Greywacke	Basalt
Vall. temple								
Causeway	X							
Pyr.temple	X	X		X		X		X
Pyramid	X	X						
Substructure		X		X			X	
Sahure	Loc. limest.	Fine limest.	Calc.alab.	Red granite	Granodiorite	Quarzite	Greywacke	Basalt
Vall. temple	X	X		X				X
Causeway	X	X						
Pyr.temple	X	X	X	X	X			X
Pyramid	X							
Substructure		X		X			X	
Neferirkare	Loc. limest.	Fine limest.	Calc.alab.	Red granite	Granodiorite	Quarzite	Greywacke	Basalt
Vall. temple								
Causeway								
Pyr.temple			X					
Pyramid	X	X		X				
Substructure								
Niuserre	Loc. limest.	Fine limest.	Calc.alab.	Red granite	Granodiorite	Quarzite	Greywacke	Basalt
Vall. temple		X		X				X
Causeway	X	X						X
Pyr.temple	X	X		X		X		X
Pyramid								
Substructure		X		X				

Djedkare Is.	Loc. limest	Fine limest.	Calc.alab.	Red granite	Granodiorite	Quartzite	Greywacke	Basalt
Vall. temple								
Causeway		X						X
Pyr.temple	X	X	X	X		X		X
Pyramid								
Substructure		X		X				
Unas	Loc. limest.	Fine limest.	Calc.alab.	Red granite	Granodiorite	Quarzite	Greywacke	Basalt
Vall. temple	X	X		X				
Causeway		X						
Pyr.temple			X					
Pyramid	X	X						
Substructure	X	X	X				X	
Teti	Loc. limest.	Fine limest.	Calc.alab.	Red granite	Granodiorite	Quartzite	Greywacke	Basalt
Vall. temple								
Causeway								
Pyr.temple		X	X	X		X		
Pyramid	X	X						X
Substructure	X	X		X			X	
Pepi I	Loc. limest.	Fine limest.	Calc.alab.	Red granite	Granodiorite	Quartzite	Greywacke	Basalt
Vall. temple								
Causeway								
Pyr.temple	X	X	X	X				
Pyramid	X	X						
Substructure	X	X	X	X			X	
Merenre	Loc.limest.	Fine limest.	Calc.alab.	Red granite	Granodiorite	Quartzite	Greywacke	Basalt
Vall. temple								
Causeway								
Pyr.temple	X	X		X				
Pyramid	X	X						
Substructure	X	X		X			X	
Pepi II	Loc. limest.	Fine limest.	Calc.alab.	Red granite	Granodiorite	Quartzite	Greywacke	Basalt
Vall. temple								
Causeway	X	X				X		
Pyr.temple	X	X	X	X		X	X	
Pyramid								
Substructure		X	X	X			X	

Tab. 2: List of rock samples taken from the various Old Kingdom pyramid complexes.

Symbol	Pharao	Dyn.	Locality	Limestone	Others	No.	Total
D	Djoser	3	Saqqara-N	67	mortar	2	69
St	Sekhemkhet	3	Saqqara-N	26			26
Ca	Khaba (?)	3	Z. el-Aryan	11			11
Na	Nebka	4	Z. el-Aryan	8			8
Sm	Huni?, Snofru	3+4	Meidum	68	mortar	1	69
Sk	Snofru Bent	4	Dahshur	76	mortar	1	77
Skk	Snofru cult py.	4	Dahshur	14			14
Sr	Snofru Red	4	Dahshur	32	mortar	3	35
C	Khufu	4	Gizeh	180	mortar	1	185
					basalt	4	
Cf	Khafre	4	Gizeh	72	mortar	1	73
R	Djedefre	4	Abu Roash	29	mortar	1	30
M	Menkaure	4	Gizeh	48	granite	2	50
Sf	Shepseskaf	4	Saqqara-S	28	greywacke	4	35
					granite	3	
Uk	Userkaf	5	Saqqara-N	42	mortar	2	49
					basalt	5	
Ni	Neferirkare	5	Abusir	27	mortar	1	30
					basalt	1	
					quartzite	1	
Cs	Khentkaus	5	Abusir	15			15
Nf	Neferefre	5	Abusir	13			13
Nu	Niuserre	5	Abusir	32	mortar	1	38
					basalt	2	
					quartzite	2	
					calcite	1	
Ub	Menkauhor ?	5 ?	Saqqara-N	10			10
SUk	Userkaf (Sun Temple)	5	Abu Gurob	9	mortar	1	15
					greywacke	3	
					quartzite	2	
SNu	Niuserre (Sun Temple)	5	Abu Gurob	18	quartzite	2	23
					greywacke	1	
					calcite	2	
Dj	Djedkare Isesi	5	Saqqara-S	31	quartzite	2	35
					calcite	2	
Un	Unas	5	Saqqara-N	33	mortar	3	45
					quartzite	4	
					calcite	5	
T	Teti	6	Saqqara-N	49	mortar	1	55
					granite	2	
					quartzite	1	
					calcite	2	
Ti	Teti Iput+Khuit	6	Saqqara-N	16			16
P I.	Pepi I	6	Saqqara-S	31	basalt	1	35
					calcite	3	
Mr	Merenre	6	Saqqara-S	26	quartzite	1	27
P II.	Pepi II	6	Saqqara-S	36	mortar	2	43
					basalt	3	
					quartzite	1	
					calcite	1	
Total				1088		88	1176

I. The Step Pyramid of Djoser (2650–2630 BC)

Geographical position (centre of pyramid): 29°52'16" N, 31°12'59" E

Height: 60 m | Base lines: 121 x 109 m | Enclosure wall: 545 x 277 m, 10.5 m high

In general, the pyramid complex of Djoser in Saqqara (Fig. 2) is regarded as the first and oldest monumental building made from formatted stones in Egypt[5]. This technical innovation, replacing the thus far used sun-dried mud bricks, goes back, according to written tradition, to the wise Imhotep, the architect of the Djoser complex. In later times, Imhotep was therefore highly honoured, in the New Kingdom even deified, and in Greco-Roman times adored as the divinity Asklepios[6]. In later periods, Djoser received the attribute "the one who opens the stone", probably in

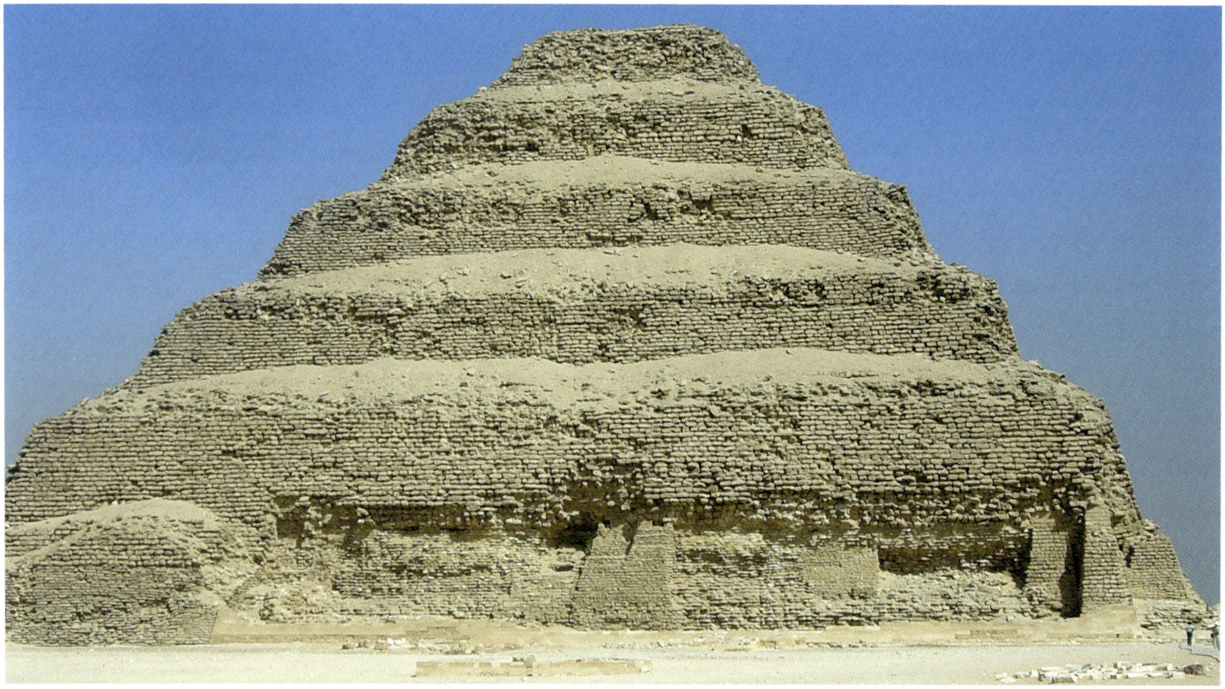

Fig. 2: Step Pyramid of Djoser at the Saqqara plateau, seen from south.

memory of the first systematic quarry work accomplished with his buildings[7]. Contrary to later pyramids, the superstructures of the Djoser complex are mainly composed of limestone blocks.

A thorough investigation of the Saqqara area was undertaken to find the origin of this limestone material, which was extracted in considerable quantities. Hereby it was presupposed that, because of these voluminous rock quantities, the production sites might not occur too far away from the construction site, in order to keep logistical problems to a minimum.

The whole mortuary complex of Djoser consists of the pyramid, built in six steps, surrounded by several architectural elements within a rectangular enclosure wall of substantial dimensions (about 1.6 km long), which shows niched panels and a total of 15 gates, only one of which was used as an entrance. Apart from the pyramid, the most striking features are two wide courtyards to the north and south of the pyramid, which remained free of building constructions. All of these elements developed in several phases over a time span of more than twenty years[8].

The royal tomb itself was originally planned as a mastaba and only later converted into a step pyramid with a mortuary temple to the north. A second tomb was constructed on the southern edge of the precinct. The burial chambers of both tombs are made of granite blocks, constructed at the bottom of deep shafts. Because of its internal and aboveground structures, only the northern tomb is regarded as the king's burial place. The meaning of the second tomb is still not clearly understood. The complex as a whole is regarded as a representation of the land of Egypt in the afterlife. The northern part containing the pyramid represents Lower Egypt with the royal palace, the courtyard with the eastern buildings the land of Egypt with its shrines, and the southern edge with the second tomb the holy site of Abydos with its cemetery of early kings[9].

In contrast to the surrounding monuments, which are built from well-dressed limestone blocks of different dimensions, the pyramid itself shows much smaller formats resembling usual mud brick sizes. This phenomenon led to the common opinion that the Step Pyramid represents the first example for the conversion of hitherto practised mud brick architecture into stone (Fig. 3). Systematic investigation of the rock formations, which delivered the limestone blocks for the pyramid's core material, gives evidence that this is a wrong assumption, as shown below in more detail.

A causeway, leading up to the south-eastern corner of the pyramid with a branch to the entrance, is supposed to have served as a construction ramp. It can be clearly seen on aerial photographs. Later, its lower part was reused by Unas as a causeway. Whether a valley temple connected to a harbour at the foot of the escarpment existed, seems doubtful, as the Step Pyramid is north-south oriented in contrast to later east-west oriented pyramids, which could be directly connected with a valley temple and a harbour east of the building[10].

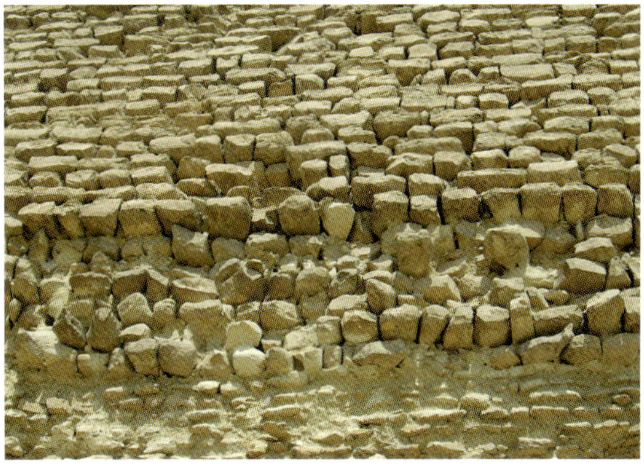

Fig. 3: Step Pyramid with small mud brick-likestone formats within the first mastaba construction phase (lower part). Note the larger formats of the later phases (upper part).

Geological environment of the Djoser complex

HUME (1911) considered the limestones of the Saqqara plateau to be of Eocene age and named the sequence simply Saqqara limestones, whereas BLANCKENHORN[11] assigned the marly, sandy limestones and marls from Abusir to Dahshur to his "untere Mokattam Stufe" (lower Mokattam formation) of the Middle Eocene.

According to the "Geological Map of Greater Cairo Area" of the Geological Survey of Egypt[12], all architectural monuments of the Saqqara pyramid field are situated geologically on the Middle Eocene Ravine Beds of the Fayum facies. This area encompasses the sun temples north of Abusir and the pyramid complex of Pepi II in the south, thus including the Djoser complex. This facies is the eastern continuation of a sedimentation succession, which occurs in its characteristic appearance more to the west in the Fayum area. The easternmost sections of this facies still occur in the group of hills of the Housan Megalli, east of the Nile, and south-southeast from where they border a distinct fault of the Wadi Hof formation of the facies. This fault, which was created as a result of the extensive tectonic events in the Miocene (the later Pliocene deposits were not influenced by this), runs in a north-northwest direction and determines the Nile valley course here. The actual Ravine Beds are described in the geological map legend as a fossil-poor series consisting of marl and marly limestones with clay interbeds.

According to the "Geological Map of Egypt, scale 1:500 000" of the Egyptian General Petroleum Corporation[13], the Djoser complex is positioned upon the shallow marine dense-bedded limestones of the Observatory (?) formation, belonging to the Mokattam Group of the Upper Middle Eocene. With its shallow marine sandy limestones, partly marly and intercalated by shale, however, it resembles rather the Wadi Rayan formation, occurring more to the south in the Dahshur area.

Our field observations and petrographic-microscopical investigations corroborate, however, the publication of the Egyptian research group of YOUSSEF ET AL.[14], whose results are based on a detailed study of the Saqqara area including stratigraphy, micro-palaeontology, and micro-facies. These authors consider the limestones of the Saqqara area as an own member within the Maadi formation of the Upper Eocene. Further, the authors present a detailed geological map including convincing fault systems, from which one fault causes the escarpment of the Abusir-Saqqara plateau, which slopes down to the Nile valley plain with a bank height of about 17 m.

The "pyramid quarries" at Saqqara

It should, firstly, be noted that the construction material for the core masonry of almost all the stone pyramids comes from the immediate vicinity of the monuments (Fig. 4). Thus, the absence of large, clearly recognizable quarry areas at Saqqara is all the more remarkable. In fact, there are obvious dug-in trench structures near the Djoser site, some of which have well-smoothed, artificially worn limestone walls, so that they can be regarded as supply areas for the construction of the pyramids and mastabas. Parts of the trench through which the Unas causeway cuts can be considered a former quarry. Nevertheless, these depressions in the ground are entirely insufficient in volume to provide the required amounts of material for all the constructions that took place at the necropolis of Saqqara. If we look further at the area for possible quarry sites, we must consider the unnaturally steep, craggy escarpment at the east of the Saqqara plateau. YOUSSEF ET AL.[15] suggest a fault between the small, sandy plain at the foot of the escarpment and the agricultural land. However, large quantities of angular rock debris were found at the foot of this plateau. This debris is typical of lime-

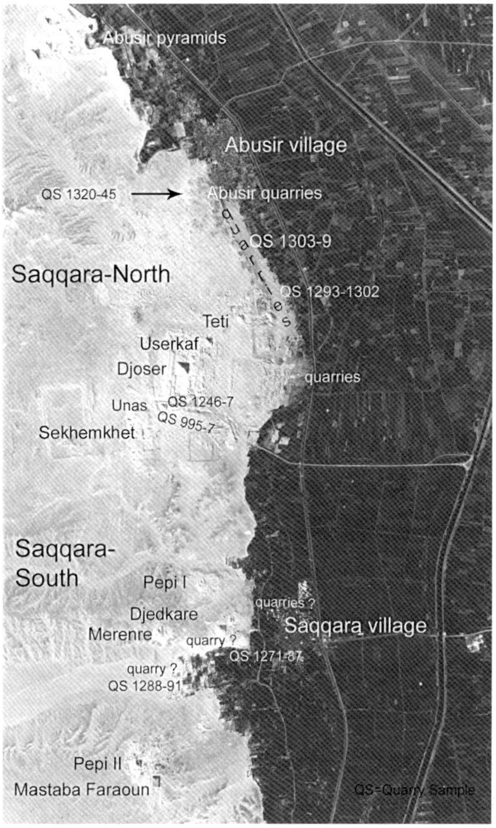

Fig. 4: Aerial photograph of the Saqqara district with the pyramids and the sampled quarry positions.

stone quarry rubble heaps in Egypt. Furthermore, rocky outcrops are discovered at many sites, particularly north of the Emery House, that had been processed in a manner consistent with quarrying. This leads to the conclusion that the steep escarpment can be thought of as an artificial wall, a relic of previous quarrying activity. In addition, the lithologically processed TM-satellite image (Fig. 1) gives clear evidence that between the actual escarpment edge and the agricultural land an intensively worked zone existed during the past.

The reason for the lack of clearly definable quarry areas, so characteristic elsewhere in Egypt, seems to be the layered structure of the limestones at this location. These layers consist of 20–60 cm thick sand-rich calcareous beds, alternated with layers of calcareous marl and clay. Due to the striking structure it was possible to easily extract the sandy limestones from their parent beds in order to cut them into the desired sizes, the original thickness being reflected in the thickness of the construction stones. Thanks to this simple excavation method, large amounts of construction stone were obtained in Saqqara within a short period of time. Most of these blocks were distinctly smaller than the ones used for the Gizeh pyramids. Until now this phenomenon, which is merely the result of the natural properties of the rock, has generally been interpreted as a deliberate following of the mud brick format, particularly in the case of the Djoser pyramid complex as the oldest monumental stone construction in Egypt.

However, this rock material has a rather low resistance to weathering. This was well recognized by the ancient Egyptians and solved by covering the pyramid with a limestone casing, which is considerably more resistant to deterioration.

It seems that there were originally geological layers offering a sufficient amount of a highly weather-resistant limestone on the Saqqara plateau, which were widely used by Djoser and Sekhemkhet.

Yet, this remarkable rock material was also used later to cover the step pyramid phases of the Meidum pyramid and can also be found in the masonry covering the small pyramid at Zawiyet el-Meitin (3rd dynasty) in Middle Egypt, as comparative analyses show.

Further, such highly weather-resistant limestone could also be detected in the limestone parts of several royal mastabas of the 1st dynasty in Saqqara-North. This fact is all the more interesting, as during such early periods, when mainly sun-dried mud brick buildings were constructed, no extended transport routes existed, indicating that the source of this stone material could have been at the Saqqara plateau itself.

The voluminous construction projects of the 3rd dynasty appear to have exhausted this source of limestone so that the subsequent kings (starting with Snofru) were forced to obtain the pyramid casing material from the quarries at Tura and Maasara on the eastern bank of the Nile (see below).

Petrography of the "pyramid quarries" at Saqqara

The petrographic fabric of the rock material used for the core constructions of all pyramids at Saqqara and Abusir is characteristically bimodal. This is a consequence of the layered structure of the geological sequence of the Saqqara Member from where it derives.

A wide spectrum of sandy to marly limestones occurs within the layered sedimentary succession cropping out at the escarpment of the Saqqara plateau, both at its eastern flank towards the Nile valley and at its northern flank towards the Abusir semi-plain (Fig. 5).

1. The fresh fracture of the first type shows a grey-yellow to yellow-brown and noticeably layered sandy limestone with fine bio-fragments, a variety of which is particularly rich in fossil fragments. However, some layers are also of a fine marly limestone which only rarely contains sand grain inclusions. They are more grey-yellow in colour and have a generally much lower fossil content.

The layered nature of the sandy limestone is created by fine clay and occasionally gypsum interstratifications, and weathering of the rock initially starts along these seams. Different qualities of this sandy limestone were used in building the pyramids, regardless of their weathering characteristics.

With a hand lens, a calcite carpet in the cavities of partly dissolved shells can clearly be seen. Further, within this type of sandy limestone, it is sometimes possible to make out a definite silification in the intergranular structure; this gives the rock a very dense consistency, which is, however, by no means more resistant to weathering.

Under microscope, these sandy limestones have a microsparitic carbonate groundmass with, in part, a relatively high (up to 5 %) distribution of limonite, giving the stone's yellowish to brownish colour. The most noticeable characteristic structural feature is the grains of quartz sand, which vary between 0.05–0.1 mm in average size. They are mostly angular and only occasionally rounded, and they can comprise up to 30 % of the structure.

Remains of fossils (globigerinae, gastropods, shell remains, and sea urchin spine fragments) dissolved in fine crystalline sparite occur up to 15 % (mostly less) of the groundmass and are generally so intensively affected by diagenesis that a precise palaeontological identification is almost impossible.

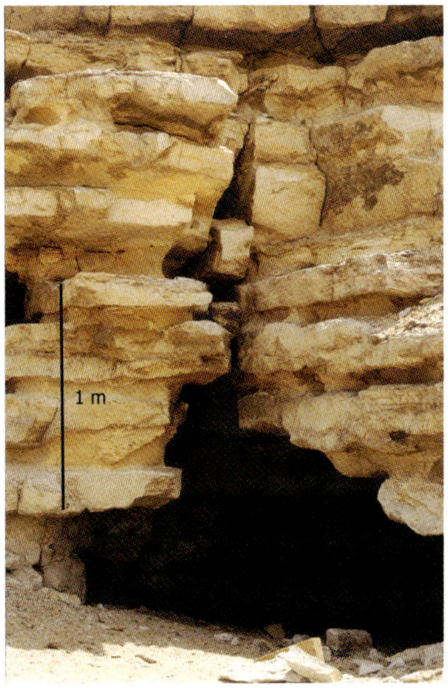

Fig. 5: Typical escarpment quarry at Saqqara-North above the Islamic cemetery of Abusir village. Note the intensive layering of the sand-rich Saqqara limestone formation yielding only small block formats.

2. The second type of limestone occurring within the layered sequence of the Saqqara escarpment is predominantly marly. These layers have a clear stratification caused by fine embedded clay layers, partly also with gypsum.

Macroscopically, the specimens show negligible fossil content. Secondary calcification is seen less often, due to the reduced ability of aqueous solutions to pass through the dense, argillaceous structure. This means that these rock qualities are considerably more susceptible to weathering. Apparently, the ancient builders were aware of this, because they used this type of stone only when there was an intermediate intergrowth with the actual calcareous sandstone.

With a hand lens, no additional or noteworthy structural characteristics can be seen.

Under microscope, the structure of these marly limestones is dominated by yellowish clay minerals, which constitute the groundmass of the rock, along with fine sparitic calcite. The groundmass also contains very fine, rough-cornered quartz grains of less than 0.05 mm in diameter. The fossil rubble with remains of globigerinae and sponge needles, found primarily in the groundmass, is in most cases partly dissolved diagenetically and can no longer be identified by palaeontologists.

The stratigraphic position of the sandy and marly limestone sequences of the Saqqara plateau are, according to the 1:100 000 Geological Map of Greater Cairo[16] the Ravine Beds (marls and marly limestone with clay interbeds) of the Middle Eocene, in the Fayum facies. According to the 1:500 000 scaled geological map of Egypt[17], these sequences belong to the Wadi Rayan formation of the Mokattam group (Middle Eocene). Intensive studies of YOUSSEF ET AL.[18] describe these series in detail and classify them as upper Calcareous Beds of the Saqqara Member of the Maadi formation (Late Eocene).

The source of this sandy limestone material is predominantly the escarpment of the Saqqara plateau, both towards the river Nile plain and along the northern edge of the plateau, but also along parts of the trench structures around the Djoser site and of the Unas causeway.

Petrographical provenance determination of the building stones

Core material

It seems that all varieties of sandy to marly limestone and calcareous sandstone described above were used for the construction of the Djoser pyramid in all its different stages.

A grey-beige, fine sandy limestone dominates a (rarer) light brownish type of similar composition, normally with a distinct layering of a few mm and almost free of macroscopically recognizable fossil remains. The pyramid stones of sandy limestone are of relatively small size with a thickness varying between 15–25 cm.

With a hand lens, they appear dense and sandy, again with no visible fossil remains. The more marly types feel less sandy, and they are of distinct bedding in mm-scale, again without any visible fossil remains.

Under microscope, sharp-edged sand grains with diameters of 0.1–0.02 mm (see the lower grain sizes below) are most striking within a calcareous groundmass (Fig. 6).

At higher magnification, the groundmass shows a mixture of clay-rich micritic and fine-grained remains of different fossils such as sponge and echinoderm spins, globigerinae and bryozoans. Fig. 7 demonstrates for comparison a thin section of the corresponding quarry material. The close similarity is striking.

The more marly limestone portions show lower grain sizes (0.03–0.003 mm) and a finer mixture of clay and calcareous mud in their sandy fraction (Fig. 8). Again for comparison, Fig. 9 gives a very similar image of a quarry sample.

For a number of other building stones from the core masonry (Fig. 10), a rather precise attribution was possible to the deep trench south of the southern enclosure wall, which is provably a Djoser quarry site. The typical algae-rich sandy limestone, which is poor in fossils, was detectable only at this location (Fig. 11).

Further, calcareous sandstone, which crops out near the modern rest house, could also be recognized in the core masonry. It is a brownish to dark brownish sandstone, which occurs as banks in the sandy limestone sequence of the Saqqara plateau.

Fine limestones used for the casing, buildings, and enclosure walls of the 3rd dynasty pyramids

Very special limestone types were used for the casing of the 3rd dynasty pyramids and for lintels, door pilasters etc. of the brick mastabas of the 1st and 2nd dynasties at the Saqqara plateau. These rock qualities are highly resistant to weathering; they immediately remind one of the fine limestone qualities of Tura and Maasara, with which they are generally confused[19].

Macroscopically, they greatly differ from the Tura and Maasara limestones as they have a rather more porous, coarse-grained structure. They are also somewhat more reddish-brown when coated with a patina. They can easily be dressed and, above all, easily polished without the surface becoming too compacted, so that the whole of the natural permeability is preserved. Thus, internal salt content is washed out and transported to the surface by movements of moisture. This contributes enormously to the rock's ability to withstand weathering once it has been built into a structure.

With a hand lens, it appears as a very dense, fine-grained limestone with no detectable fossil remains.

Under microscope, these relatively fine limestones consist of a dense micritic calcite groundmass containing some extremely fine clay minerals, which give the groundmass a dark colour in microscopic sections.

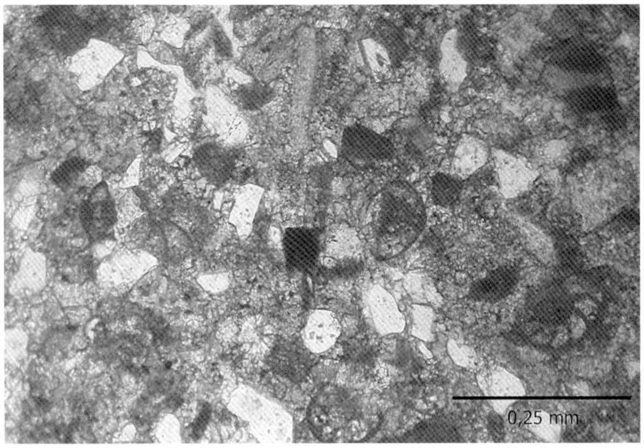

Fig. 6: Micro-photo (D 38) of sandy limestone from Djoser core material. Note the white sand grains within a micritic groundmass with remains of micro-foraminifers and sponge spins (1 pol. filter).

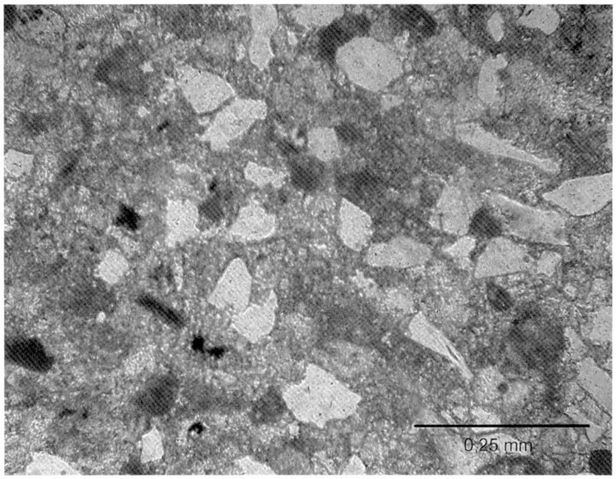

Fig. 7: Micro-photo (QS 1315) of a sandy limestone from the quarry sites of the eastern Saqqara plateau escarpment. Note the almost identical structure with the core material of Fig. 6 (1 pol. filter).

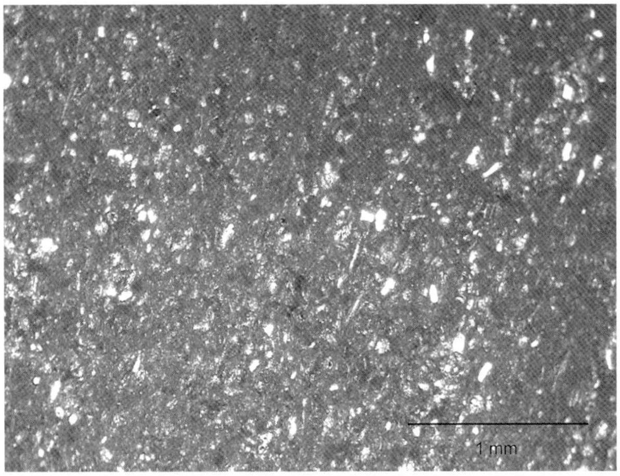

Fig. 8: Micro-photo (D 42) of fine sandy to marly limestone from the Djoser pyramid's core masonry. White sand grains in a fine micritic matrix of clay and micro-calcite with sponge needles (1 pol.filter).

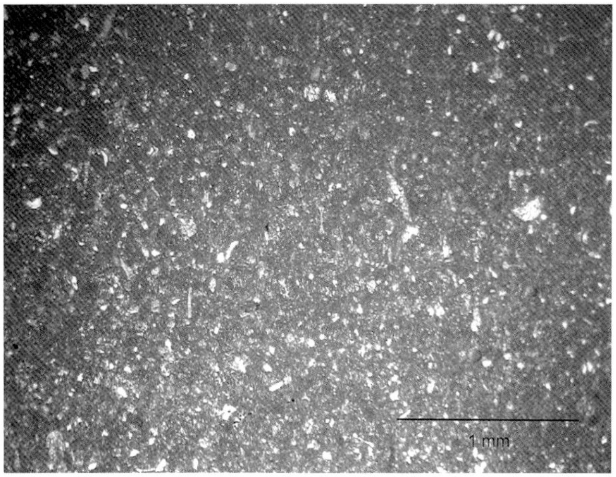

Fig. 9: Micro-photo (QS 1316) of fine sandy to marly limestone from the quarry sites of the eastern Saqqara plateau escarpment. Note the almost identical structure with the core material of Fig. 8 (1 pol. filter).

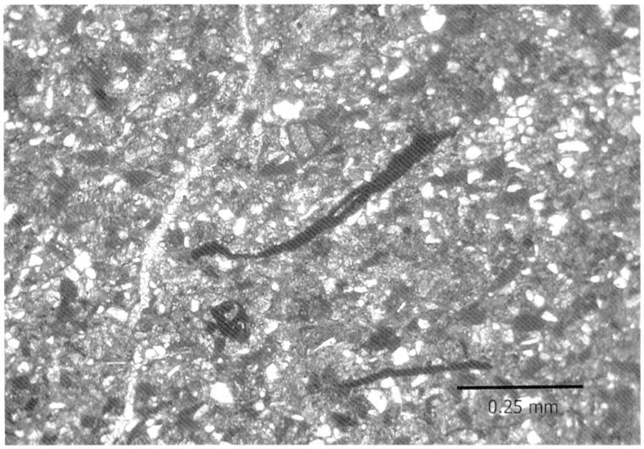

Fig. 10: Micro-photo (QS 1007) of a sandy limestone from the core masonry of Djoser pyramid. Note the sandy structure and the black filaments of bituminous matter (1 pol. filter).

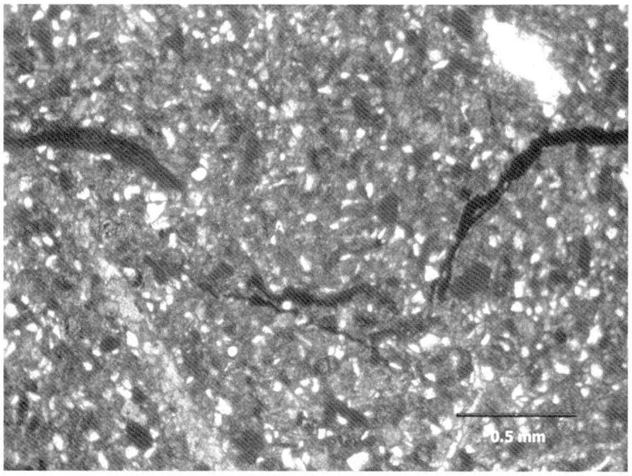

Fig. 11: Micro-photo (QS 1246) of a sandy limestone from the trench south of Djoser southern enclosure wall. Note the sandy structure and the black filaments of bituminous matter (1 pol. filter).

Within the matrix, the approx. 5–10 % fossil components stand out due to their brightness. This is caused by the slightly coarser crystallinity of the calcite within the pore space of the organogenetic structures, some of which were originally present and some of which were recrystallised by diagenetic calcification of the fossil component. This latter process dissolved most of the primary structures of the small fossils, most of which are globigerinae and rims of very small nummulites, so that only their contours can be made out in the groundmass (Fig. 12).

In spite of intensive searching, a quarry site for this limestone material could not be found at the Saqqara plateau, but fragments of this stone type occur there almost everywhere. Furthermore, architectural limestones used in mud

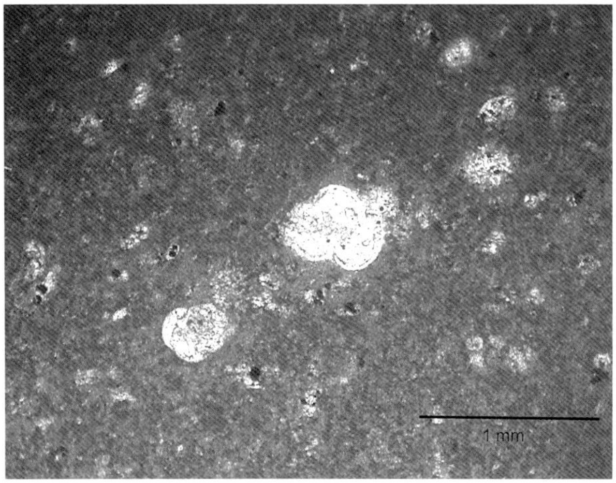

Fig. 12: Micro-photo of a casing stone (D 9) from Djoser pyramid. Within an extreme fine-grained, marly-calcareous groundmass completely calcified globigerine micro-fossils occur, but not anymore palaeontologically identifiable.

brick mastabas of the 1st dynasty in the northern parts of the Saqqara plateau are also of the same rock structure. These observations indicate that a remarkable amount of such a stone type must have occurred at the plateau, where it was exhausted already in an early stage.

As a consequence, from the reign of Snofru, who still used this rock type for the casing of his true pyramid completion of the Meidum step pyramid and also partially as casing for the Bent Pyramid, the (somewhat less weather-resistant) limestone from Tura and Maasara became the main source for the casing material of the pyramids.

Stratigraphic position: Despite the intensive recrystallization, M. BOUKHARY from Ain Shams University Cairo (personal communication), who kindly restudied our sections using a special method for the extraction of nano-fossils, was able to classify this fine limestone according to its planktonic foraminifera, nano-plankton and a number of ostracodes to the Bartonian age (Uppermost Eocene). This would fit into the upper continuation of the Saqqara Member stratigraphy.

It is a disadvantage of the present study that the original quarries of this special stone quality could not be identified, despite intensive field investigations on the Saqqara plateau. Thus, we assume that a limestone layer which originally existed in the area could have been completely mined out at the end of the 3rd dynasty, leaving behind only traces of waste. However, the casing stones not only from the Djoser complex but also from the complexes attributable to successing kings of the 3rd dynasty ample sufficient evidence of it, as will be demonstrated below.

Geochemical provenance determination of the core material from the Djoser pyramid

In the binary correlation diagram of Mg versus Fe (given in ppm = parts per million), the samples of the core masonry of the Djoser pyramid (Fig. 13) show a rather dense distribution pattern plotted closely to the distribution field of the north-eastern quarry samples of the Saqqara area. Within this diagram, the data of the samples from Djoser's core masonry plot only in a relatively small range with Mg-values lower than 10 000 ppm and Fe lower than 4000 ppm. This limitation allows the conclusion that the building material was also derived from a limited area of the Saqqara-North plateau. These results, however, have to be confirmed by further geochemical diagram comparisons.

Analogous results may be seen in the correlation diagram Mg versus Sr. However, here a striking concentration of the data corresponds with the QS 1303–1309 quarry site (Fig. 14).

The geochemical data allow the provenance localization of the core material to only a relatively small and limited range within the correlation diagrams of the Saqqara quarry series, which is located in the close vicinity of the pyramid, along the nearest parts of the escarpment towards the Nile valley. These findings correspond well with the petrographical results.

Comparing the various sizes of the core material, it becomes evident that the originally planned mastaba contains stone of strikingly small size, with an average width of 60 cm, a thickness of 15–20 cm and a depth of 60–80 cm. In relation, the block formats of the later stages of the actual Step Pyramid are clearly larger. This might be explained by the progression of the quarry workings at both the plateau itself and at the escarpment. Due to the faster weathering of the portions richer in clay, the harder sandy limestone beds remained in the upper near-surface parts only as layers of minor thickness (15–20 cm), because the almost fully deteriorated interim material disintegrates to debris, when worked. When further quarry progress comes into less deteriorated parts of the layered rocks, larger block formats could be obtained. The less resistant character of the lime-clay interlayer was either not registered or simply regarded as less

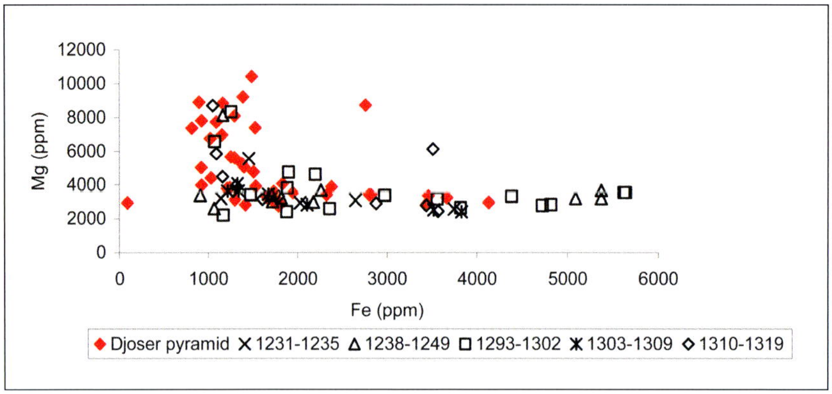

Fig. 13: Mg/Fe-diagram from Djoser pyramid's core material and sandy limestones from the Saqqara-North quarry sites. The close correspondence of core and quarry material is obvious.

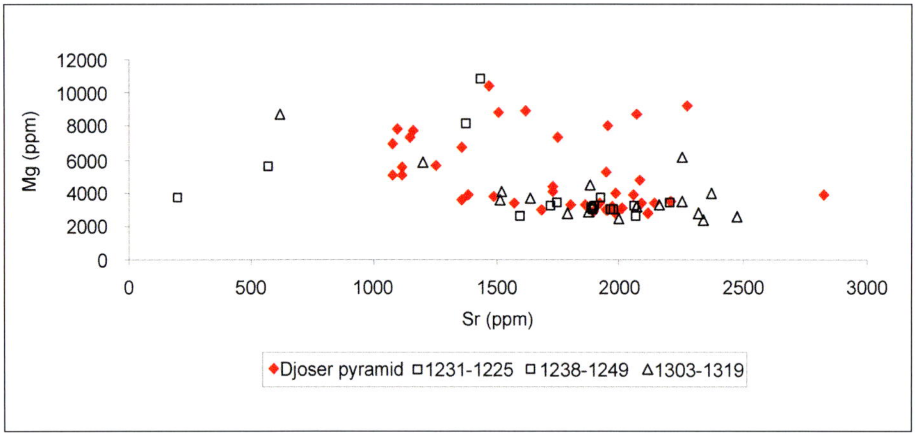

Fig. 14: Mg/Sr-diagram from Djoser pyramid's core and sandy limestones from Saqqara-North quarry sites. Note the close correspondence of core and quarry material, also in this diagram version.

important for the stability of the construction, as they were protected by the up to 6 m thick wall of backing and casing blocks. After the removal of this highly weather-resistant protection cover in later (Arab?) times, the deterioration of these larger building stones was relatively faster than that of the mastaba stage, today causing an impression of wider inter-block joints.

Geochemical provenance determination of the casing material from the Djoser pyramid and the surrounding complex

The petrographic description of the stone material used in the entire Djoser complex already shows a clear bimodality. The relative soft stone material of both the pyramid's core masonry with its low resistance to deterioration and, as far as exposed, of the inner parts of the wall stands in contrast to the casing of all the various constructions within the surrounding complex. This, of course, was always recognized by archaeologists working in that area[20] and in analogy to the Gizeh pyramids, this casing material is generally classified as a rock type deriving from Tura[21].

Petrographically, it has already been clarified that this limestone, due to both its fossil record and its internal structure, could by no means be confused with the limestone of the areas of Tura, Maasara or even of Mokattam. But the globigerina dominated fossil content as well as the facies characteristics in combination with BOUKHARY's fine stratigraphic attribution into the Uppermost Eocene (Bartonian), gives a hint to the Saqqara plateau as a possible source.

On the other hand, in spite of that, the limestone qualities of the quarry site south of the Meidum pyramid fit with their fossil content – and to some extent also geochemically – into the frame of this casing limestone, thus making this quarry site a second provenance candidate. Whatever, the close similarity of the casing material of all three early pyramids (Djoser, Sekhemkhet, and Meidum) makes the probability of a mutual source very likely.

In Fig. 15, the Mg/Fe-data for the casing of the three early pyramids plot into an almost identical distribution field, probably indicating a mutual source area.

Fig. 16 shows comparable information, whereby the values of the casing stones of both the Djoser and Sekhemkhet pyramids can be plotted into a very similar distribution field together with the samples of the quarry to the south of the Meidum pyramid, again indicating a close geochemical relationship and fitting with the above-mentioned petrographical results.

Studying Fig. 17, a clear differentiation of the quarry samples from the area south of the Meidum pyramid and the casing material from the three pyramids is obvious. Only two quarry samples correspond with this latter group (Fig. 18). This may indicate that they derive from a limestone bank which was almost completely mined out in ancient times. This would explain the limited availability of this highly weather-resistant limestone, i.e. for only the three oldest pyramids. In regards to later constructions, the search for new casing limestone sources was inevitable.

On the other hand, the far higher Fe- and Mn-contents of the limestones from the quarry south of Meidum (Fig. 17), which are only comparable to the core limestones of the Meidum pyramid, contradict the close geochemical correspondence of the casing material with the samples of the Mg/Fe- and Mg/Sr-data from the area to the south of Meidum. This indicates that the casing stones possibly originated from the latter site.

An additional interesting differentiation can be seen in the Fe/Mn-diagram of the casing stones (Fig. 18). Further examination of the fields of low Mn-content between 20 and 70 ppm establishes that here predominantly the samples of the primary mastaba are plotted, which were taken at the southern side of the pyramid. The samples with a higher Mn-content between 90 and 130 ppm derive mainly from the casing, from the buildings, and enclosure wall of the Djoser complex. A similar, but less striking bimodality can also be seen in the Mg/Fe- and Mg/Sr-diagrams (see Figs. 15–16). The differences, however, are too small to assume a different provenance. This may indicate a slight geochemical variation laterally within one bank of a geological unit. Such geochemical variations along lateral extensions of stratigraphic limestone strata with identical facies lateral could be observed repeatedly during earlier investigations

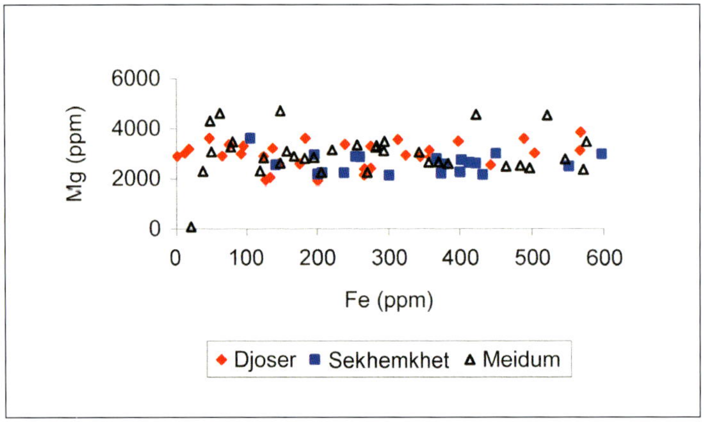

Fig. 15: Mg/Fe-diagram of casing stones from Step Pyramid and mortuary district in Saqqara together with casing stones from Sekhemkhet's unfinished pyramid and casing stones from Meidum pyramid. The close correspondence of all three limestones within a well-defined distribution field is obvious.

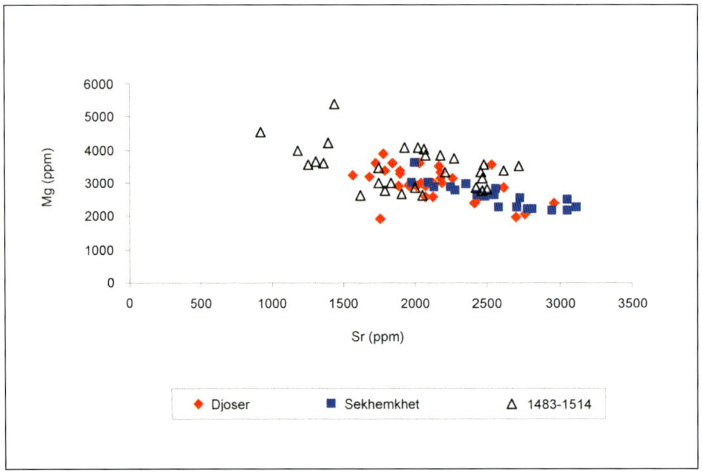

Fig. 16: Mg/Sr-diagram of casing stones from Step Pyramid and mortuary district in Saqqara, casing stones from Sekhemkhet's unfinished pyramid and samples from the quarry area south of Meidum pyramid.

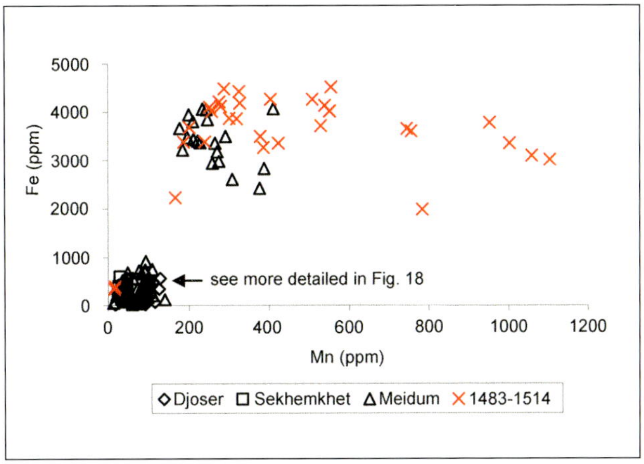

Fig. 17: Fe/Mn-diagram of casing stones from Djoser pyramid and his mortuary district in Saqqara, from Sekhemkhet and Meidum pyramids, and the entire samples from the quarry south of Meidum. Clearly recognizable is the good correspondence between the casing stones of the Meidum pyramid and the quarry stones, but also a significant difference to the various casing stones in a separated field with far lower Fe- and Mn-values.

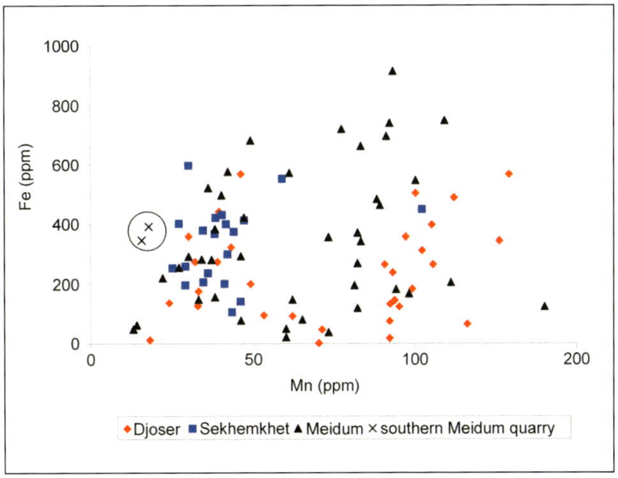

Fig. 18: Fe/Mn-diagram of casing stones from Djoser pyramid and his mortuary district in Saqqara and from Sekhemkhet and Meidum pyramids. Due to the small Mn-scale of this diagram, the casing stones of all three pyramids are better separated but plot closely within a limited field. Note that only two samples of the quarry material plot within this low Fe/Mn-field, indicating that only a small amount of this limestone remains in the present day quarry site.

of ancient Egyptian limestone provinces[22]. In other words, this is not a speciality of the Meidum or Saqqara area. Most probably during the different stages of construction activity, the fine casing limestones were mined in quarry areas with slight lateral geochemical variations.

In contrast to the distribution fields of the geochemical analysis of the core material within the various diagram types, the casing samples give a distinctly closer distribution pattern and far lower Mg- and Fe-contents. This causes, from a geochemical point of view, a higher resistance to weathering (lower Mg-content) and further, a brighter white-grey colour (lower Fe-content), which changes over the centuries to a slightly yellow-brownish tinted patina.

II. The "Buried Pyramid" of Sekhemkhet (2630–2625 BC)

Geographical position (end of descending corridor): 29°51'59" N, 31°12'47" E

Estimated height: ~ 70 m | Present height: 8 m | Base: 120 m | Slope: ~ 72°

The unfinished pyramid of Sekhemkhet shows some particularities, which are worth mentioning here. Originally planned as a step pyramid with presumably seven steps, it remained unfinished except for the first step and part of the second, which were both covered by sand until their discovery in the mid 20th century[23]. In its unfinished state, the pyramid's height is about 7 m, and one may calculate a total height of the finished construction of about 70 m, which means 10 m higher than the Djoser pyramid. The block sizes of the remaining superstructure are bigger than those of the Djoser pyramid, but are in a similar way slightly inclined towards the centre. An outer casing had either not yet been installed or removed at a later stage. From the north, an entrance passage slopes down under the middle of the pyramid where the burial chamber and adjacent galleries can be found. An alabaster (calcite-alabaster) sarcophagus with a unique vertically sliding door at one end was found empty[24].

The whole complex shows in its original plan striking similarities to the Djoser complex, not only in its dimensions, but also in the whole layout with its north-south orientation, separate south tomb, wide courtyard in the north, and niched enclosure wall with niche panels. Nevertheless, it remained unfinished, probably because of the short 6 years' reign of the king.

The close vicinity of the Sekhemkhet pyramid to the Djoser complex also results in a similar lithological source of both core and casing material. However, the quarry area of this pyramid is most probably situated relatively close to the south-western area of its base. This would explain, in spite of the many lithological similarities of the core and casing material, some of the remarkable differences to that of the Djoser complex.

Due to the unfinished state, an intensively worked quarry area cannot be expected and is all the more unlikely as such an artificial depression at the Saqqara plateau would in the meanwhile most certainly be heavily sand blown.

Core material

Not very much of the core masonry is exposed at the present state of this mostly unexcavated pyramid complex (Fig. 19). Thus, the rock sample variation is less representative in relation to the samples taken at the Djoser complex. The relatively poorly formatted stone blocks of local origin were generally bound with a coarse mortar.

Macroscopically, the main type immediately recalls the marly limestone of the Djoser complex. It is light grey-beige and rarely yellowish in colour, and feels much less sandy than the second type (see below). It shows only a very fine layering with an occasionally disorderly sedimentation.

With a hand lens, these limestones appear dense; the very small-sized sand grains are mostly invisible, but on sedimentary plains small mica grains reflect the sunlight. No fossils or fossil remains are detectable with this method.

Under microscope, a dense micritic groundmass, which contains many fine fragments of mainly sponge needles and globigerinae, dominates the entire structure. Rarely, slightly dissolved globigerinae still exist and allow a stratigraphic attribution. The extremely fine clay minerals are hard to identify under the microscope and darken the thin section to a brownish-grey tint. The sand grains are mostly less than 0.01 mm in diameter and consist generally of quartz, seldom plagioclase grains (Fig. 20).

A second type of limestone is used to a lesser degree in the masonry; a yellow to almost brownish sandy limestone, occasionally grading into lime-rich arkosic sandstone. In hand, they feel somewhat sandy but not to such an extent as a true sandstone. This is all the more apparent, as characteristic bubbles appear when a drop of dissolved hydrochloric acid is applied. In spite of the many micro- and nano-fossils, no macro-fossils can be detected. Ripple marks and fine cross patterns could be recognized within the well-developed sedimentary fabric.

Under microscope, the many sandy silica components dominate the rock fabric. The sharp-edged grains consist predominantly of quartz, but also of feldspar (plagioclase and microcline) and rarely hornblende, biotite, tourmaline, and zircon. Generally, the sand grains are on average 0.2 mm in size. The groundmass is mainly micritic to partly oolitic, intermixed by fine milled fossil debris, which is impossible to define more precisely, but with relics of globigerinae and sponge needles.

The third type, again light grey to slightly beige is a very fine-grained compact limestone, which differs macroscopically from the casing stone only by its less whitish colour.

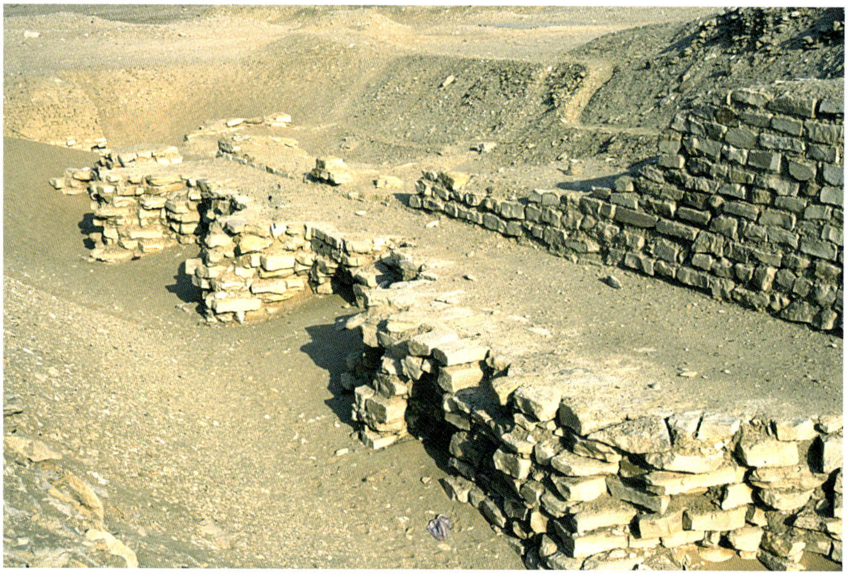

Fig. 19: Sekhemkhet's unfinished pyramid in Saqqara, north-west corner. Remains of magazines surrounded the pyramid's first step (right).

Fig. 20: Micro-photo (St 15) of core limestone of Sekhemkhet's pyramid. The oval darker portions within the matrix of this rather fine-grained, sandy limestone are former peloids, the entire fossil content in the groundmass is destroyed by diagenesis (1 pol. filter).

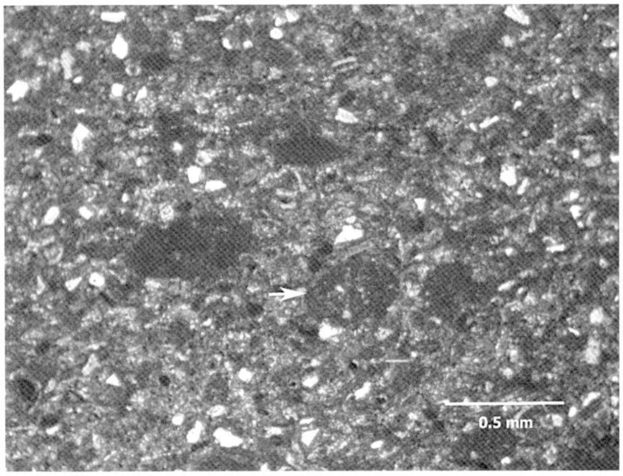

Under microscope, the entire micritic groundmass consists only of partly dissolved relicts of globigerinae and sponge needles. Secondarily calcified globigerinae rarely occur as light objects in the matrix, which is darkly coloured by extremely fine clay.

This type seems to form a link with the fine whitish-grey limestone of the casing material.

Casing material

Casing material from the Sekhemkhet pyramid complex is at its present state mainly restricted to the "White Wall" in the north of the pyramid[25], most probably part of an originally planned enclosure wall. Its large, right-angled blocks were carefully smoothed in the same way as those of the respective wall of the Djoser complex.

The strikingly white colour of this limestone material deserves explanation.

In contrast to the casing used for the Djoser complex, which appears yellow-reddish in colour, here the casing stones are of a bright whitish-grey colour. This variation in colour is easily explained by the circulation of even small amounts of moisture in the dry desert environment during the many centuries of exposure. This brings low amounts of iron-containing solutions towards the surface, where they then oxidize to a brownish-red three-valence iron oxide. As Sekhemkhet's complex lay buried at least since New Kingdom times, such a surface oxidation process could not take place. Consequently, the casing material has remained whitish-grey. The entire Djoser complex must have appeared as bright during ancient periods.

However, the very close relationship to the casing material of the Djoser complex will be proven below geochemically.

Under microscope, this very fine-grained dense micritic limestone is of a rather brownish tint due to its relatively high content of sub-microscopic clay minerals. Within this matrix, up to 5 % is composed of relicts of former globigerinae and sponge needles, which appear light-coloured because of a diagenetic calcification of their pore volumes.

Due to the very close similarities of this casing limestone to that of the Djoser complex they both should be of the same stratigraphic relations.

Geochemical provenance determination

The existing pyramid base obviously consists purely of core masonry, and we could not find any unambiguous relicts of original casing masonry at the site.

The block sizes are comparable to those of the later construction phases of the Djoser pyramid but clearly larger than those of its primary mastaba (see Fig. 19). An explanation of the increase in block size has already been offered above, in connection with the Djoser pyramid. However, the predisposition of the stone blocks to deteriorate, as mentioned above, could not be observed at the Sekhemkhet complex, due to the better conservation of the blocks beneath the debris and sand coverage. This might also be the reason that the block sizes appear to be larger than of the later phases of the Djoser pyramid.

The geochemical correlation diagram (Fig. 21) shows a distribution pattern for the core masonry of the Sekhemkhet pyramid, which corresponds well with the Saqqara-North quarries, but obviously with distinct variations. Therefore, and due to the close petrographical relationship, we may postulate that the quarry area of Sekhemkhet's pyramid belongs to the same geological unit as that of the Djoser complex, and indeed, it occurs within the plateau, which is the western continuation of the escarpment. The Mg/Sr-diagram (Fig. 22) shows very similar results.

The casing material was sampled at the northern white enclosure wall (Fig. 23). The data of the samples plot almost entirely into the separated field of the Djoser casing blocks (Fig. 24) with lower Mn-values. The close concordance of the Sekhemkhet and Djoser casing stones is also evident in the geochemical correlation diagram Mg/Sr (Fig. 25). This provenance may be explained by the close accessibility for the casing of both the Djoser and Sekhemkhet pyramids. This might possibly give hints for a more precise localization of this enigmatic limestone layer, which was used exclusively for the casing of the 3rd dynasty pyramid constructions. Again, the very close similarity of the geochemical data of the Meidum casing stones should be emphasised (see Fig. 15).

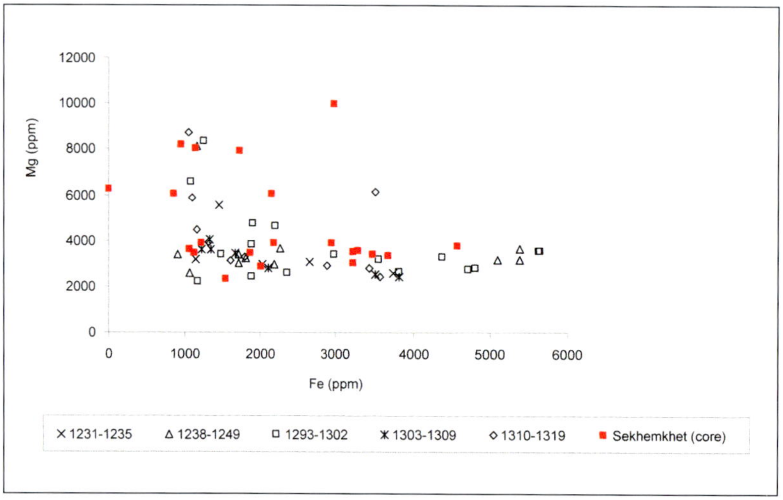

Fig. 21: Mg/Fe-diagram of the Sekhemkhet pyramid's core material and Saqqara-North quarries.

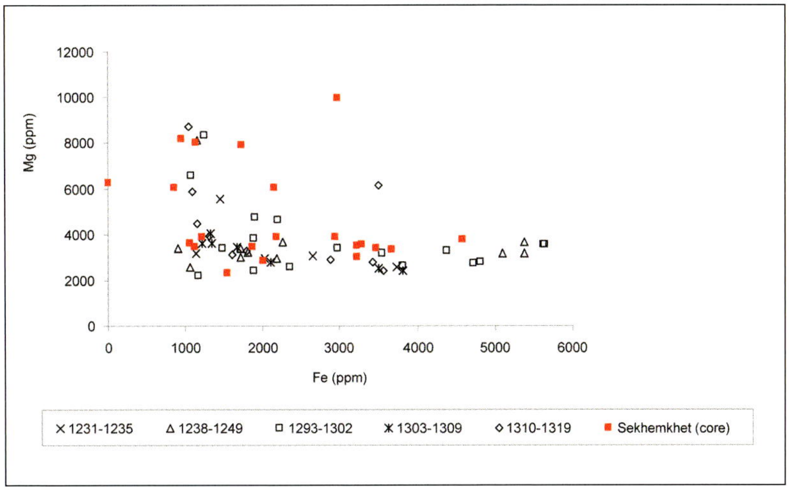

Fig. 22: Mg/Sr-diagram of the Sekhemkhet pyramid's core material and Saqqara-North quarries.

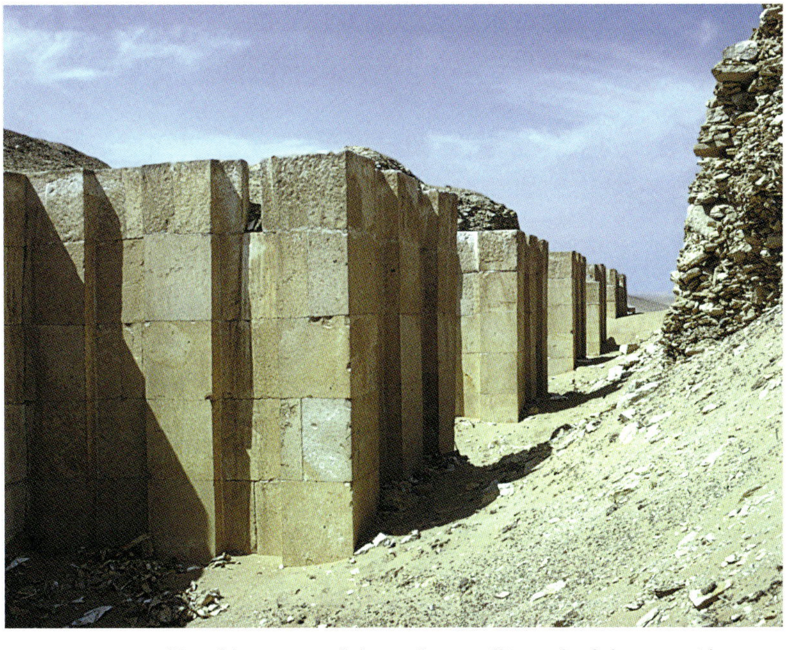

Fig. 23: Sekhemkhet. View of the "White Wall" north of the pyramid. It is built from large, carefully smoothed blocks of whitish limestone.

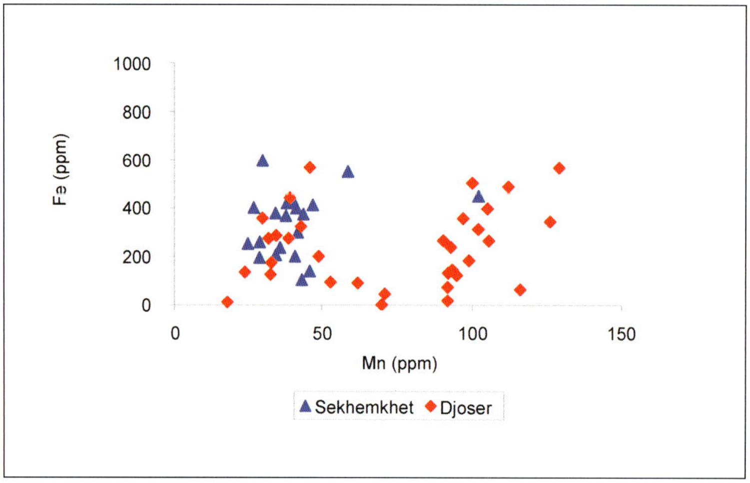

Fig. 24: Fe/Mn-diagram of casing stones from the white enclosure wall of Sekhemkhet pyramid district and from Djoser pyramid. The data plot mainly in a manganese-restricted field of Djoser casing limestones.

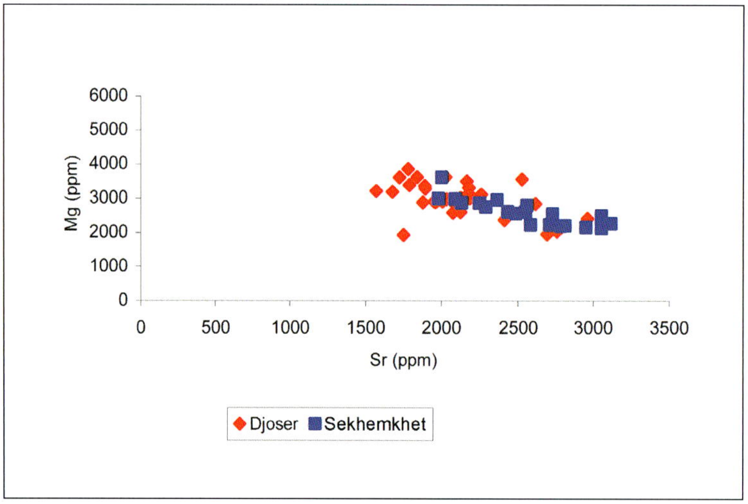

Fig. 25: Mg/Sr-diagram of casing stones from the enclosure wall of Sekhemkhet pyramid district. The Sr-values plot slightly increased only between about 2 000 and 3 000 ppm, yet generally close to the comparable distribution field of Djoser.

III. The Pyramids of Zawiyet el-Aryan

1. Layer Pyramid of Khaba (?) (2625–2600 BC)

Geographical position (centre of pyramid): 29°55'58" N, 31°09'40" E

Estimated height: ~42–45 m | Present height: ~ 17 m | Base: 84 m | Slope: ~ 70° | Inward inclination of the stone layers: 22°

Mastaba Z-500 (Fig. 26), situated north of the pyramid (at Zawiyet el-Aryan), contained eight alabaster vessels with the name of Khaba, a king at the end of the 3rd dynasty. Thus, the pyramid was attributed to this king by REISNER/FISHER[26], which is generally accepted by Egyptologists today. This attribution is supported by the construction style of the pyramid, which was obviously designed as a step pyramid with slightly inclined layers of relatively small stones of local material, recalling the pyramids of Djoser and Sekhemkhet. Casing blocks, if ever applied, are no longer present. Most probably, the pyramid was never finished.

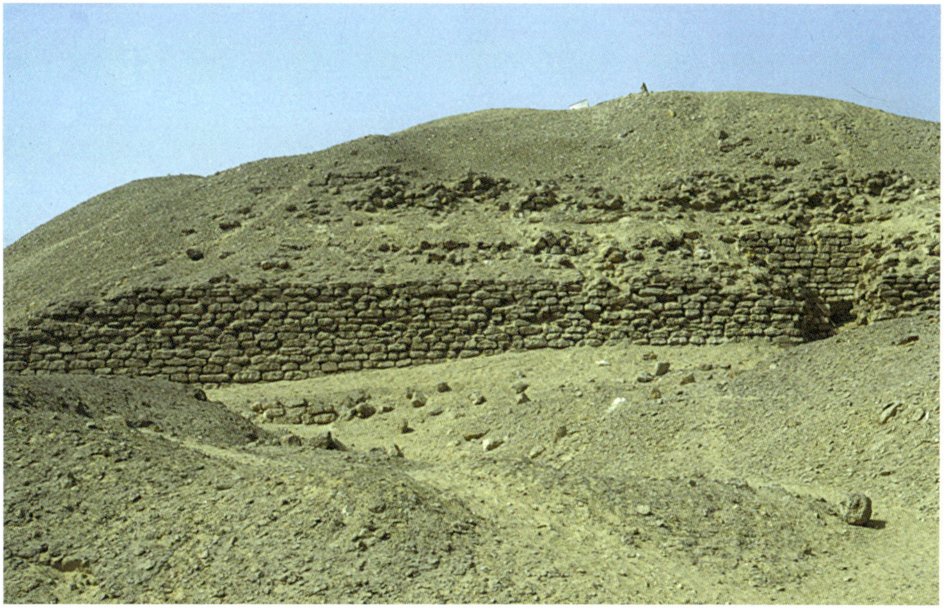

Fig. 26: Layer Pyramid (Khaba?) near Zawiyet el-Aryan, seen from east.

Apart from the early excavations of REISNER/FISHER[27], the pyramid has never really been investigated in detail. A first topographic map of the site was published only recently[28].

To the east of the pyramid, on the edge of the escarpment, a valley temple may have been associated with the pyramid and would have been the first valley temple oriented east-west. This, however, has not yet been confirmed by archaeological investigation.

In this area east of the pyramid stand the remains of brick walls that may have originally belonged to mastabas and/or possibly even a mortuary temple. The function of these ruined buildings cannot as yet be determined.

The entrance to the pyramid substructure is located near the northeast corner of the pyramid, in a trench in which a steep staircase leads to a westward oriented corridor and finally into a vertical shaft. At the base of the shaft, there is a corridor with around 32 storage annexes intended for the burial equipment.

Another corridor leads directly to a burial chamber that is located exactly beneath the vertical axis of the pyramid[29].

Archaeological investigation of this pyramid is seriously hindered due to the occupation of the entire area of Zawiyet el-Aryan by the Egyptian army.

Core material

The different varieties of the local stone material used for the core masonry can be easily studied, as later a trench was cut through the pyramid right into its centre.

The size of the local building stones varies in length between approx. 80–100 cm, in width between 40–60 cm and in height between 15–20 cm (max.). They are bound with a thick mortar filling of clayish *tafla*, which is not an ideal mortar due to its low binding strength.

The building stones were procured about 200 m north-east of the pyramid at a sharp-edged sandy and fossil-rich limestone bank, which consists of different limestone layers. The lower layer is olive green to partly brownish and

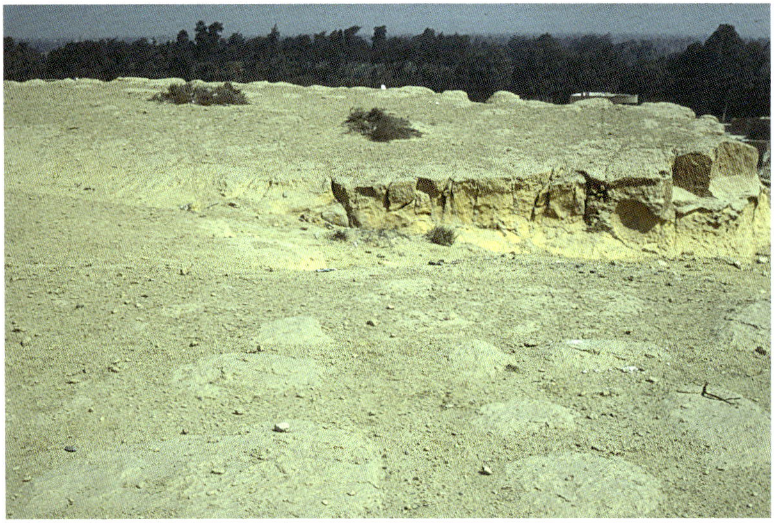

Fig. 27: Remains of the quarry site about 200 m east of the Layer Pyramid. Well recognizable are the weathered structures of a former block-wise worked quarry in the foreground and on the plateau behind.

the upper one is a brownish, hard limestone layer of about 1–1.5 m thickness. The building stones of the above-mentioned sizes were extracted from both layers (Fig. 27).

Yet, a possible quarry site, shown with a question mark in the map supplied by LEHNER[30], located in the area to the south-west of the pyramid could not be confirmed during our quarry survey.

Stratigraphically, these limestone qualities correspond to the Saqqara Member of the late Eocene Maadi formation.

2. *Unfinished pyramid of Baka (?) (2546–2496 BC)*

Geographical position (burial chamber): 29°56'23" N, 31°09'23" E

Base: ~200 m | Unfinished

The pyramid of Baka exists only in its very first stage of construction. It is situated about 150 m from the edge of the escarpment at Zawiyet el-Aryan, next to the modern asphalt road. It was investigated at the beginning of the 20th century by BARSANTI[31].

The owner of this pyramid is still under discussion, but the similarities to the 4th dynasty pyramids, especially those in Abu Roash, strengthen the assumption that it was built under Baka, the short-reigning successor of Khafre in the 4th dynasty[32].

Only a sloping corridor remains, which leads to a shaft about 21 m deep and of dimensions of approx. 12 × 25 m. The bottom of the shaft is paved with huge blocks of granite and limestone. Both the corridor and the shaft of the burial chamber were cut into the natural bedrock (Fig. 28). On the western side of the burial chamber, there was a massive oval granite sarcophagus sunk into the floor. Although the cover was intact, the sarcophagus was empty. Further, the foundations of a trapezoid enclo-

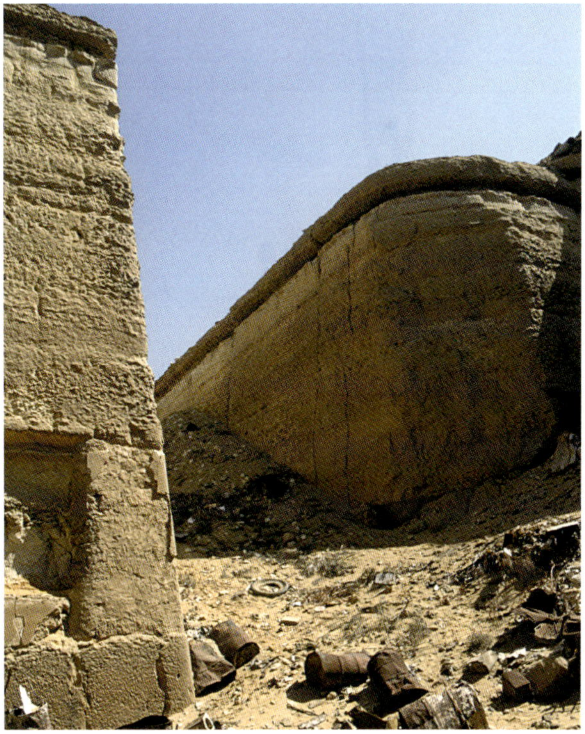

Fig. 28: Unfinished pyramid at Zawiyet el-Aryan. View from the burial chamber towards the north into the inclined corridor, cut into the natural rock.

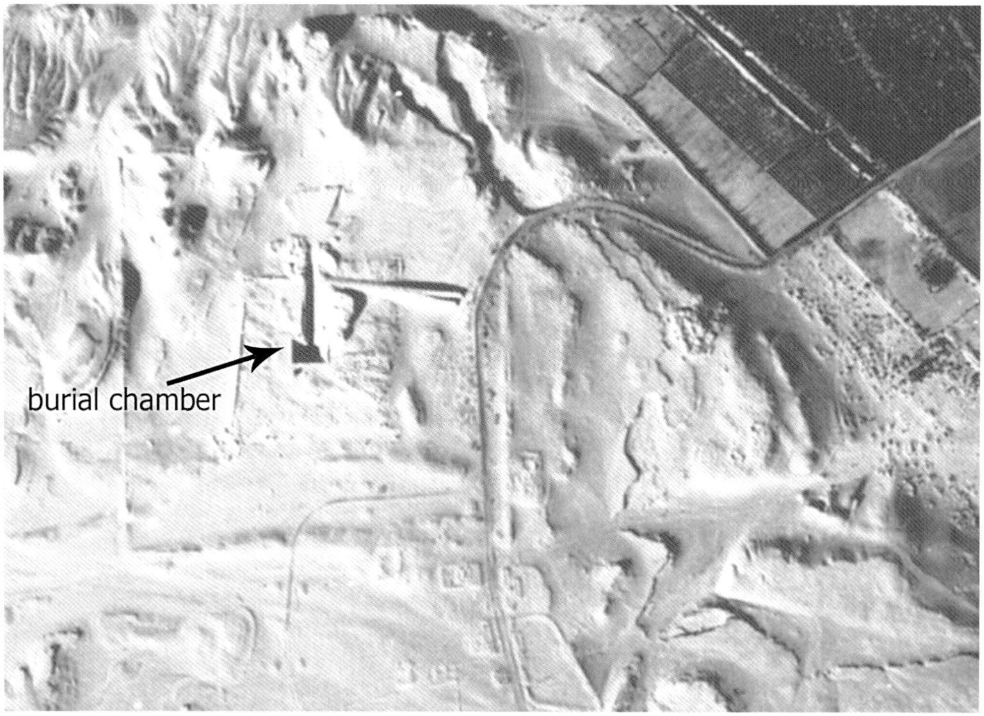

Fig. 29: Aerial photograph from about 1975 of the unfinished pyramid district at Zawiyet el-Aryan and environment. Note the unexcavated mastabas next to the enclosure wall (possibly containing the pharaohs identity).

sure wall still exist. Large limestone blocks, obtained from the excavation of the corridor and shaft at the beginning of the 19th century, are orderly deposited within the enclosure wall, forming a kind of stone magazine. The blocks are strikingly larger than those of the Djoser or Sekhemkhet pyramids, but smaller than the Gizeh pyramid blocks. They are identical with the geological rock units of the excavated corridor and burial chamber, geologically belonging to the Saqqara Member of the Upper Eocene according to YOUSSEF ET AL.[33].

However, some of the blocks of the magazine around the shaft and the burial chamber are of a sandy, brownish limestone, rich in fossils such as oysters. This material does not derive directly from the excavated burial chamber and entrance corridor of the site, but from the nearby group of hills, and belongs geologically to the Kom el-Shellul formation of the early marine Pliocene. These units will be discussed later in more detail in connection with the Red Pyramid and the Mastabat el-Faraoun.

Numerous granite blocks with traces of more modern wedge splitting together with the entire layout of the interrupted working activities recall the unfinished pyramid construction at Abu Roash. Yet, the fascination of Baka's pyramid results from its exemplification as a pyramid construction in its earliest stages.

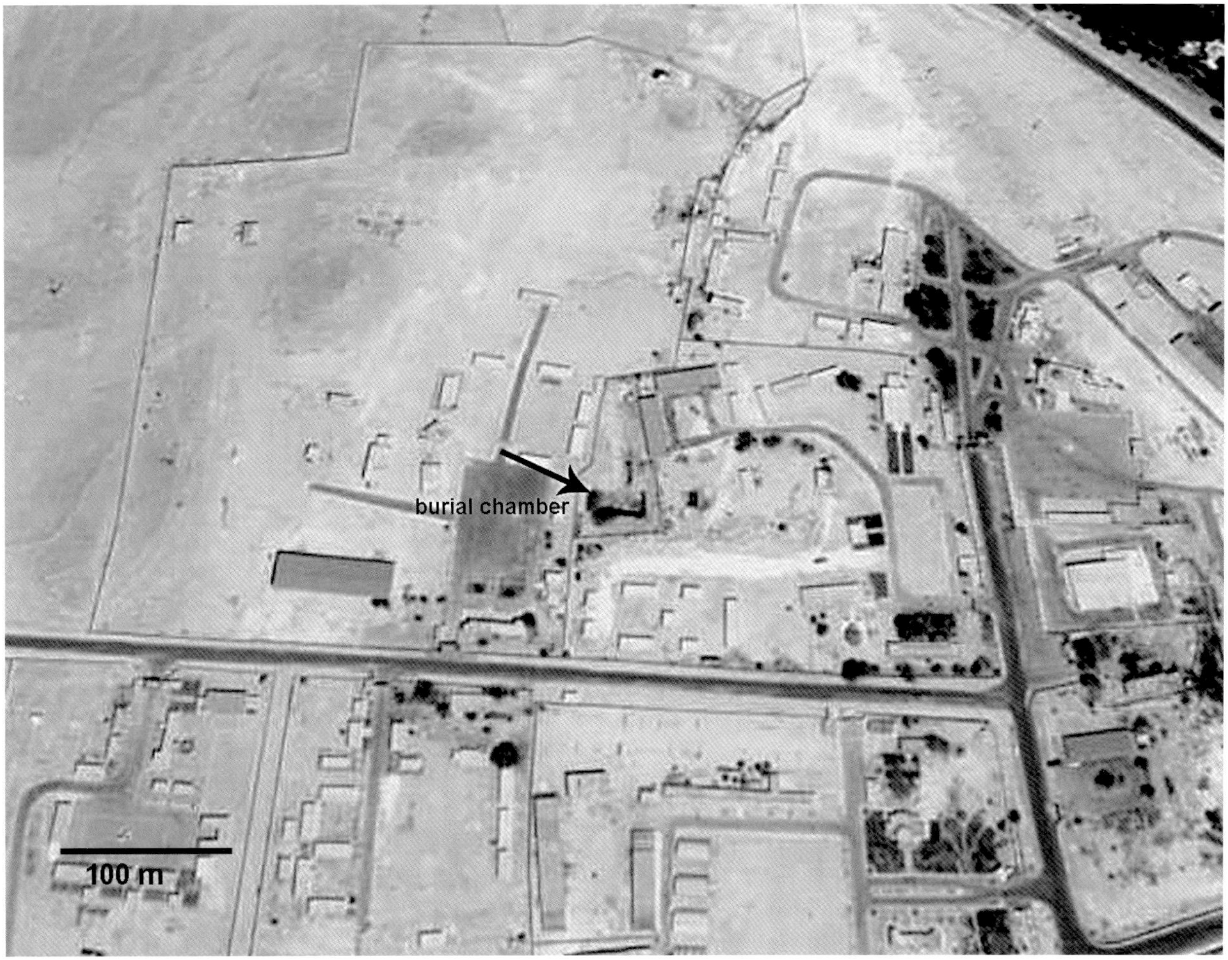

Fig. 30: Google image from 2004, showing about the same area as Fig. 29. The intensive overbuilding of the entire archaeological area except the small part of the pyramid's inclined shaft with burial chamber becomes evident.

A comparison of an aerial photograph from 1975 (Fig. 29) with a Google satellite image from 2004 (Fig. 30) clearly shows an intensive modern day overbuilding of the entire archaeological complex at Zawiyet el-Aryan. The position of the ancient enclosure wall and at least two neighbouring mastabas can no longer be made out.

IV. The Meidum Pyramid

Geographical position (centre of pyramid): 29°23'18" N, 31°09'28" E

Original height: ~92 m | Base: ~144 m | Slope: 51°

East of the Fayum Oasis, at the edge of the escarpment of the Western Desert, are the strange looking remains of a pyramid (Fig. 31). In its present stage, it is about 65 m high. Due to its extraordinary shape, compared with all the other Old Kingdom stone pyramids, the locals call it "el-Haram el-Kaddab", the "false pyramid". The name of its owner is still disputed among Egyptologists.

Due to its original design as a step pyramid, which greatly resembles the small step pyramids that are widespread in Egypt[34] and are unequivocally dedicated to Huni, the last king of the 3rd dynasty, the Meidum pyramid was ana-

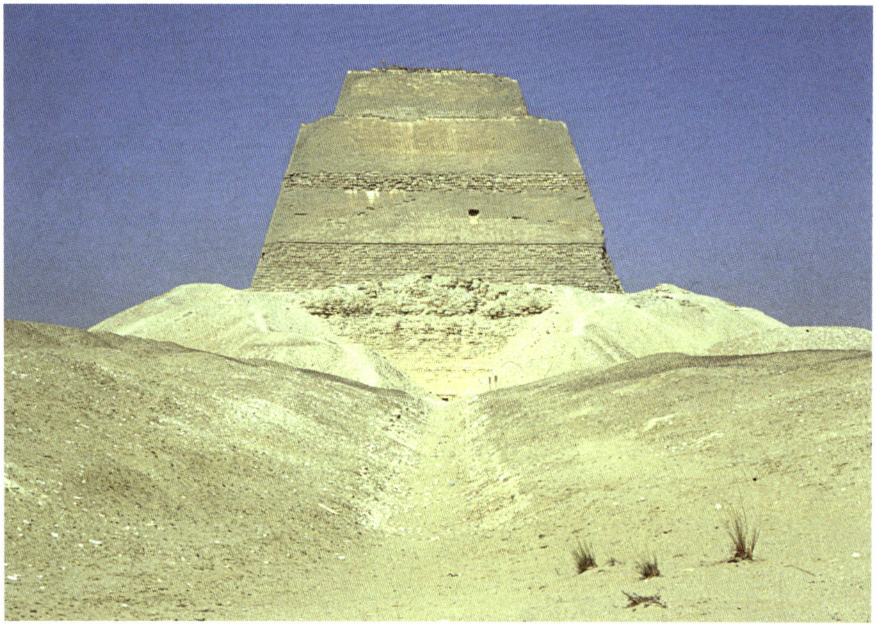

Fig. 31: Meidum pyramid with its characteristic tower-like original structure, seen from the east with causeway and small temple at the end.

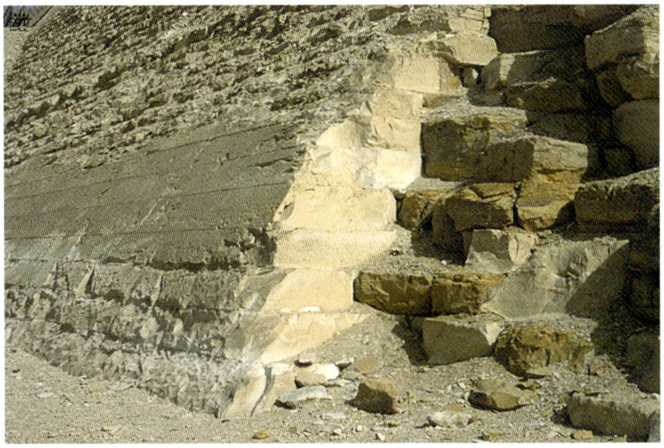

Fig. 32: Meidum pyramid with its lower part of construction phase E 3 (after BORCHARDT 1928). Note the coarse core masonry (right) and remains of the polished fine casing of the true pyramid stage.

Fig. 33: Meidum pyramid, western side. Note the change from the inclined layering of construction stage E 2 (left) to the horizontal layers of the true pyramid stage E 3 (right). ▶

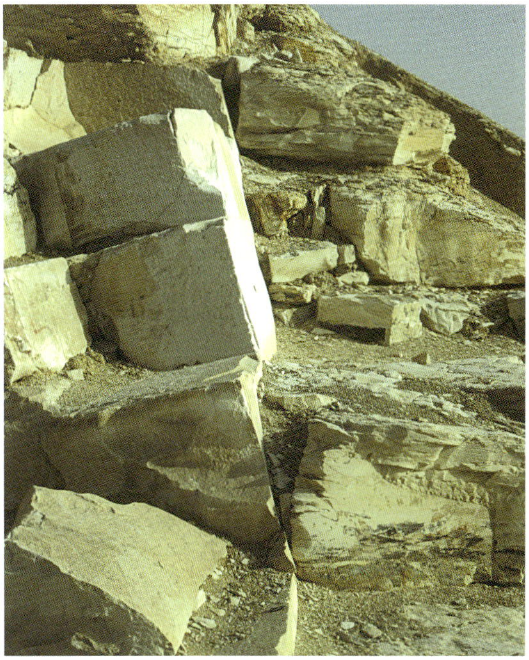

logously attributed to him[35]. Only the transition from the step pyramid to the true pyramid of the 4th dynasty was assigned to Snofru by these authors. However, Huni's name was never found at the Meidum pyramid itself or in the nearby tombs. The name of Snofru is only mentioned in visitors' marks left on the walls of the small mortuary temple east of the pyramid during the New Kingdom.

In the 1990s the large gravel mound around the pyramid was partly cleared away, whereby the polished whitish-grey casing of the last construction stage E 3 came to light[36] (Fig. 32). Further, a number of pre-prepared building stones marked with graffiti belonging to this final stage were excavated from the debris. The graffiti mention dates of years 30, 32 and 34, thus pointing towards the reign of Snofru, which possibly lasted for more than 40 years[37]. This, together with the later visitors' graffiti, convinced STADELMANN[38] to vehemently contradict the thesis of a possible double ownership of Huni and Snofru in Meidum.

After first investigations[39], it was BORCHARDT[40] who studied the Meidum pyramid intensively and established its construction history. He determined three main construction phases: E 1–3. Originally built in seven steps (E 1), it was later extended to eight steps (E 2) and finally, plans were carried out to transform it into a true pyramid (E 3). Each construction phase was cased with polished limestone, obviously planing each of these to be the final building.

Studying the archaeological situation after the removal of the debris around the north-west corner, there are some striking differences between the construction methods of the step pyramid and the later true pyramid stage. The step pyramid was constructed in line with the tradition of the previous pyramids built by Djoser, Sekhemkhet and Khaba with inclined accretion layers and, originally, a seven step structure (plus an eighth step added later) on a solid rock foundation. In contrast, the true pyramid was founded on three layers of limestone blocks, laid directly on top of the sand. Further, the blocks of the true pyramid phase (E 3) were no longer inclined towards the centre but laid horizontally as in all later pyramids from the 4th dynasty on (Fig. 33). This significant change in construction methods seems to run parallel to the change from the 3rd to 4th dynasty. Thus, from an archaeological point of view, a first and original ownership of the Meidum pyramid by Huni, the last king of the 3rd dynasty, cannot completely be brushed aside.

Petrographical provenance determination of the building stones

Systematic sampling of the various construction phases yielded all together 70 samples. It must, however, be considered that it is almost impossible to take samples from the upper parts of the remaining pyramid without expensive scaffolding, which was not available within this project.

During sampling, it became evident that most of the exposed masonry parts of the pyramid consist of casing stones and that unambiguous core material is only rarely available in the areas of the removed steps. Thus, the majority of the samples were taken from the outer casing and less from the core material.

Petrographic investigations of both core and casing material incorporated into the masonry of the Meidum pyramid clearly show that all components derive from a single distinct geological unit. Of course, there is quite a spectrum of variations, but the fossil content and the internal structure of the various rock specimens are so similar that a single provenance is unquestionable. In spite of that, the building stones vary in colour from light grey to yellow-brown. They all are slightly marly, a quality easy to test by touching the fresh surface with the tongue, which gives the feeling of adherence. Some of the building stones split into 5–7 cm large and 0.5–1 cm thick flakes when loosened

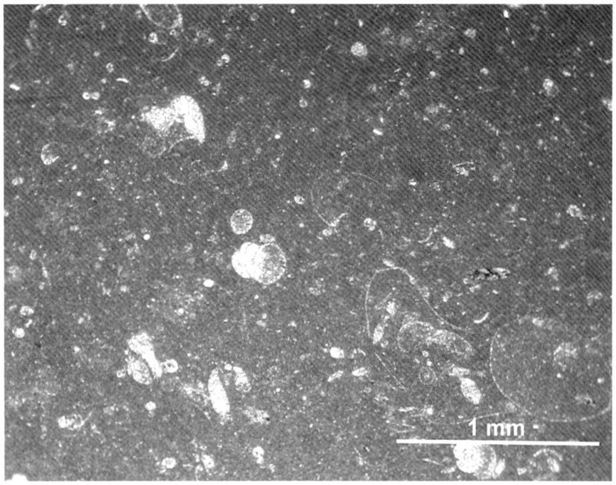

Fig. 34: Micro-photo (Sm 58) of limestone from Meidum pyramid temple. Comparing with Fig. 35, the almost same appearance turns out with dark micritic matrix and light calcified micro-fossils of mainly globigerines and remains of shells (1 pol. filter).

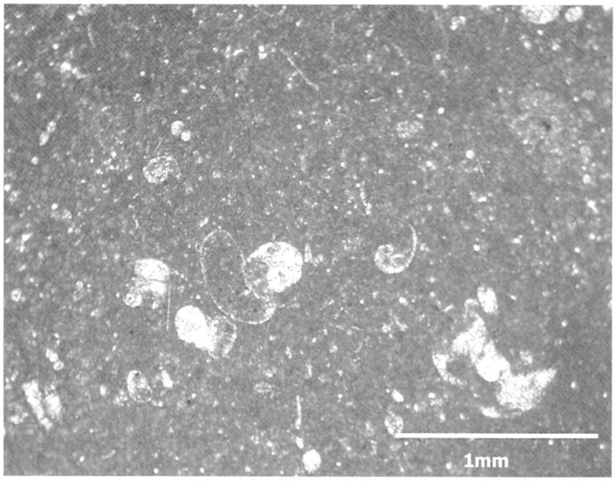

Fig. 35: Micro-photo (QS 1474) of limestone from the southern Meidum quarry with the typical dark micritic matrix and light calcified micro-fossils, mainly globigerines (1 pol. filter).

Fig. 36: Meidum pyramid. Intensive weathering into hand size chips of core blocks of construction phase E 3 (left). The white block (right) belongs to the casing of step pyramid stage E 2.

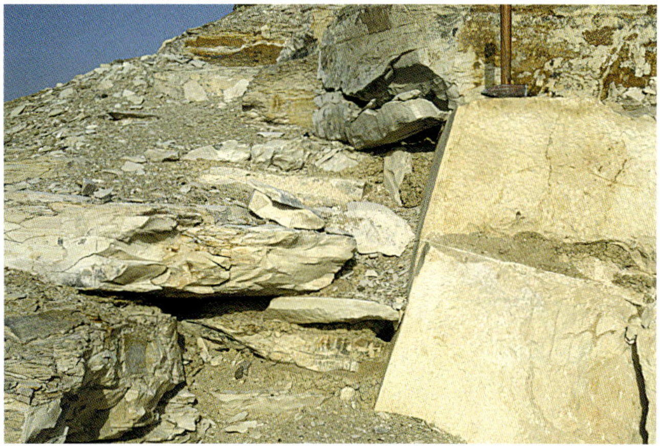

in the masonry, due to fine sedimentation bedding, which is however hard to recognize with the naked eye. Yet, most remain solid regardless of their stage of deterioration.

With a hand lens, the structure of all the samples seems very dense with no identifiable grains or fossil remains.

Under microscope, all samples are of a very fine-grained and mostly dark, dense matrix structure due to submicroscopical clay minerals. However, the nano-fossil remains appear always light, because all of them are completely overgrown by secondary calcification and thus impossible to identify palaeontologically.

Nevertheless, ghost structures of globigerinae (around 0.1 mm in size) and other foraminifers, sponge needles and finely milled fossil debris are definitely recognizable (Figs. 34–35). Only in a few cases do these fossil structures exceed more than 5 % per volume and they are always intensively calcified.

Further, especially the more yellowish-brown specimens clearly show fine striated bedding, which seems responsible for the exfoliation frequently observed in such stones exposed to weathering (Fig. 36). The more greyish varieties are normally free of any alignment and thus are highly weather-resistant when incorporated into the masonry. The latter variety is predominantly used for casing stones.

The stone varieties described above occur exclusively within the entire masonry of the construction phases E 1–2, and they dominate the unfinished E 3 phase masonry, with which the step pyramid was intended to change to a true pyramid. However, blocks from other provenance could be sampled from the casing of the E 3 phase.

Macroscopically and with a hand lens, the quality of the latter greyish variety does not differ from the relatively weather-resistant grey-whitish variety of the above-described local limestone. Under microscope, however, the structure appears very different and greatly resembles the Maasara limestone that occurs on the opposite bank of the river Nile and was used for the casing stones, especially for Snofru's Bent- and Red Pyramid (Figs. 37–38). This, of course, needs more explanation: For the casing of phase E 3, both varieties of a grey-whitish limestone were used; the dominant variety extracted from the nearby quarry site and the second type far away from the escarpment on the opposite bank of the river Nile. Therefore, it might be presumed that the nearer supply was exhausted. Consequently, a substitute had to be found, which was available only at the other, further location.

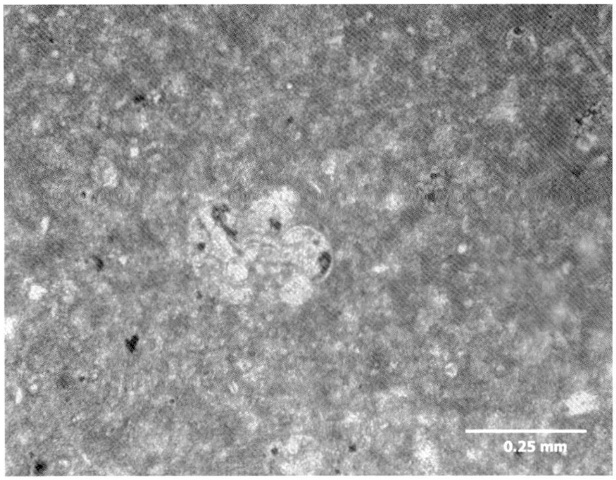

Fig. 37: Micro-photo (Sm 18) of a sample from the casing of the true pyramid construction phase E 3 of Meidum pyramid. Compare the structure and fossil remains with Fig. 38 of limestone from Maasara to realize the close similarity of both images (1 pol. filter).

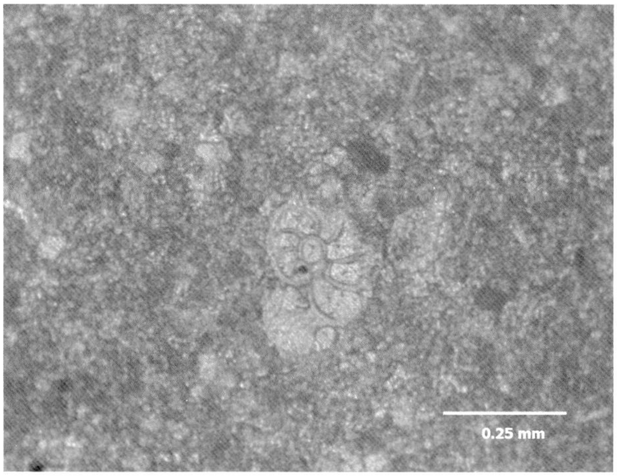

Fig. 38: Micro-photo (QS 1160) of limestone from Maasara quarry area. Compare structure and fossil content with Fig. 37 to realize the close similarity of both images (1 pol. filter).

Both the sudden change of the construction plan from a step to a true pyramid and the necessity to obtain casing material from the opposite escarpment some 25 km away may give an indication for a change of the builders personalities. It was Snofru, who first used this new limestone type for the casing of the Bent- and Red Pyramid and introduced the construction of true pyramids.

It might not be a coincidence that the small pyramid at Zawiyet el-Meitin (today Zawiyet el-Sultan) opposite Minia was covered by the same limestone type as in the phases E 1 and E 2 of Meidum. The Zawiyet el-Meitin pyramid is attributed to Huni[41]. However, due to the celebration of Horus Snofru in visitors' inscriptions dating mainly to the New Kingdom on the walls of the small mortuary temple at the eastern side of the Meidum pyramid, this king was regarded as the builder of the Meidum pyramid by others. The already mentioned consideration that Huni had originally started to build a step pyramid like his predecessors, together with the fact that after his early death, his successor was the first with the idea to build a true pyramid, can be looked at from a different perspective with the scientific results presented here. After having realized that the local casing stones could not supply the necessary demand, Snofru opened the Maasara quarries as a substitute source. Subsequently, he gave up the plan to usurp Huni's pyramid and started his own pyramid projects at Dahshur.

The quarries of Meidum

In the area around the Meidum pyramid, three sites were sampled due to their possible usage as quarries. These are, in analogy to the Saqqara quarries, firstly the less prominent outcrops along the escarpment, secondly a possible but not entirely convincing quarry site about some 600 m south of the pyramid and finally, an extended quarry area about 1 km south of the pyramid (Fig. 39).

Fig. 39: Overview of Meidum pyramid district and the nearby quarry site (modified detail of Google image 2006).

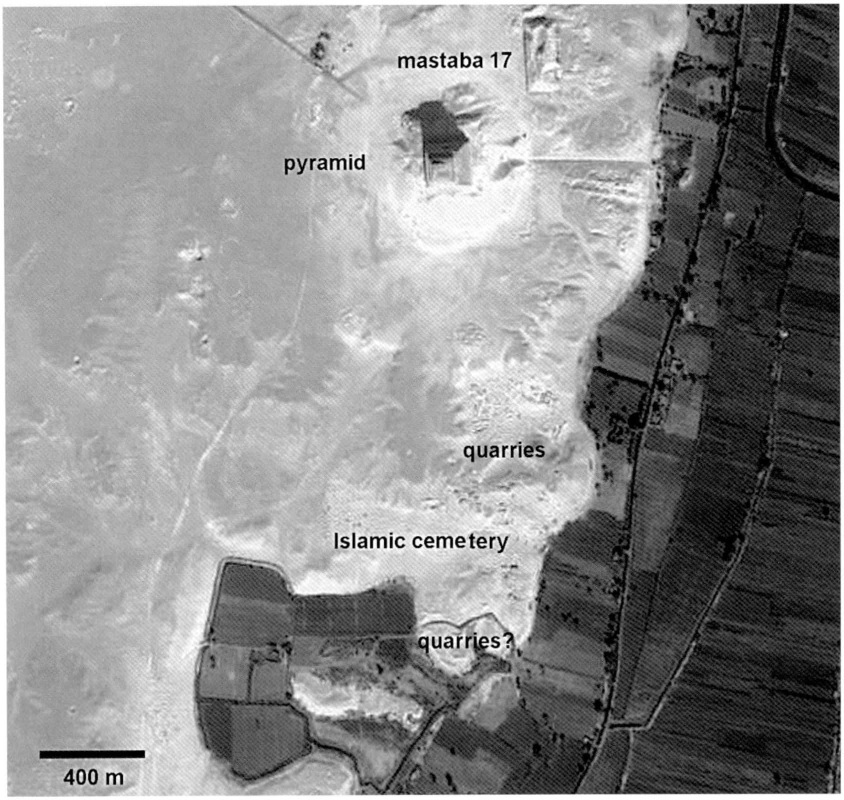

In the near vicinity of the slightly elevated site of the Meidum pyramid, no rock outcrops of a suitable building stone occur. The second possible quarry site, some 600 m south of the pyramid, which is exposed by a modern trench, did not fit to the core or casing masonry on petrographical and geochemical analysis.

Only the somewhat chaotic quarry site about 1 km further to the south, surrounded by an Islamic cemetery, is of importance for our investigation. At first glance, it appears doubtful that this site might have been used in the construction of the Meidum pyramid, particularly as it is heavily filled up with sand. A number of remaining limestone outcrops of small sizes enclose intensively worked areas. Today, it only vaguely looks like a typical ancient quarry site. However, the macroscopic appearance of the various rock qualities, both in structure and colour, seems to be identical with the building stones of the Meidum pyramid.

Altogether 31 geological samples were taken from this site, representing a good cross-section of the limestone qualities that occur here.

In general terms, there are two types of marly limestone that crop out at this site: One of a more yellowish colour, frequently with a fine laminated bedding, and in many cases strongly deteriorated and even exfoliated. Further, a second rock type of lighter grey colour appears more massive, but in exposed parts becomes brittle and disintegrates into flakes that can be up to hand size. In the case of the geological samples taken at the site, this disintegration continued years on, indicating that they were already pre-affected by weathering.

Whether macroscopically or with a hand lens, no fossil remains could be detected in the dense, fine-grained structure of the two stone types.

Under microscope, the correspondence of the rock varieties occurring in this quarry site with those incorporated into the Meidum pyramid, both in core and casing, is so evident that there are no doubts concerning the provenance attribution.

Only the more advanced stage of deterioration of the quarry stones in relation to those in the pyramid's masonry needs an explanation: Generally, building stones, whether brick or natural stone, enjoy a much better protection against atmospheric influences when incorporated into masonry than those exposed in abandoned opencast quarry sites. This is exemplified in the Gizeh quarries, where the outcropping rock units suffered more deterioration than their equivalents in the masonry of the respective buildings.

Comparing the petrography of the Meidum building stones with those of the casing of the earlier Saqqara pyramids of Djoser and Sekhemkhet, a close concordance becomes evident. The question arises as to whether the Saqqara casing stones derived from the Meidum quarries, or, if not, a comparable geological unit had once cropped out in Saqqara but had been completely exhausted, leaving only a few traces as loose blocks or flakes.

If this is so, the next pyramid-constructing pharaoh, when searching for a comparable rock type, was only successful in the Meidum region. Here, an equivalent rock quality was found, from which the inferior quality was used for core masonry and the superior quality for the outer casing of the Meidum pyramid.

No doubt, these presumptions need further investigation, especially in connection with a more intensive geological study of the limestone units along the Saqqara-Meidum sequences.

Geochemical determination of the building stones and nearby quarry sites

According to the geochemistry of the three sites, the sites along the escarpment and the one about 600 m south of the pyramid cannot be regarded as the sources of the building stones. They need no further discussion here.

The three geochemical correlation diagrams of the samples from the Meidum pyramid and the southern quarry, about 1 km south the Meidum pyramid, are somewhat confusing at first glance. However, they might be helpful in establishing the provenance of the casing material of the Djoser and Sekhemkhet monuments.

Regarding Fig. 40, two distinct geochemical distribution fields become obvious: one with Fe-values below 1000 ppm and a second field above 2000 ppm. The latter shows a close correlation with the field of the southern quarry in Meidum, even when of slightly lower Mg-content. The phenomenon that a slight geochemical variation may exist between a quarry and the stone material extracted from it, is not surprising. Only the remaining walls of the quarry could be sampled and therefore a slight geochemical variation within the earlier mined parts of the site could not be taken into consideration because these parts are now incorporated into the building.

In contrast, Fig. 41 shows a very close correlation between the pyramid materials with that of the southern quarry, indicating that during the entire sedimentation of the limestone unit no significant Sr-variation took place. This is typical for stable geological facies conditions, as already demonstrated by the petrographic record. It also indicates that the two types of building stones differ only slightly in iron content, but considerably in their deterioration behaviour.

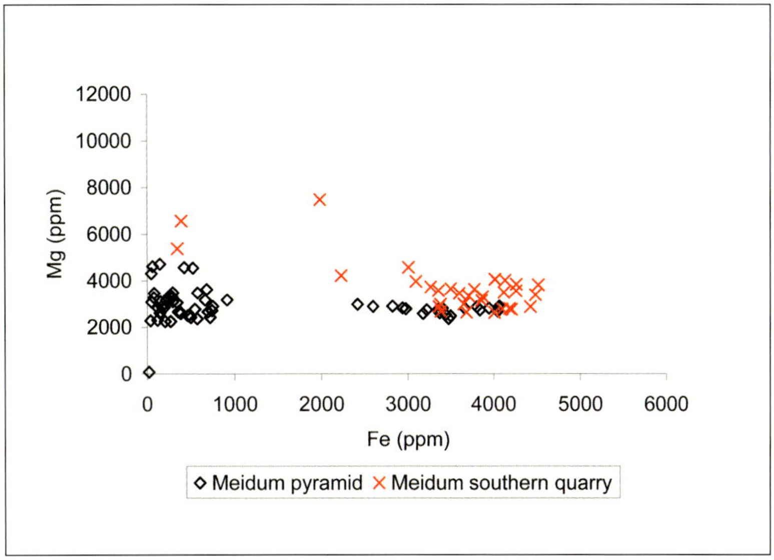

Fig. 40: Mg/Fe-diagram of the samples from Meidum pyramid and its southern quarry. The pyramid samples with Fe-values lower than 1000 ppm are casing stones. Note that only two quarry samples correspond with the casing stones, indicating that only little of this stone quality remained. Those above 2000 ppm Fe represent core samples and cluster in close vicinity with the quarry samples.

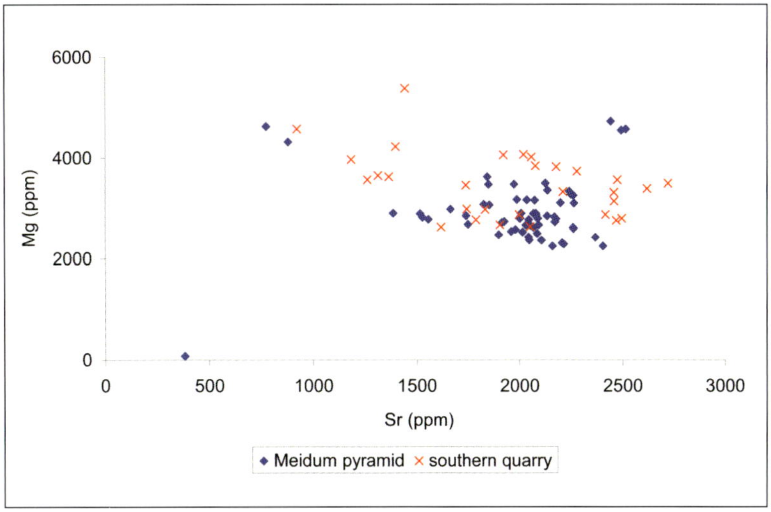

Fig. 41: Mg/Sr-diagram with a very close correlation of the entire pyramid samples with those from the southern quarry. This indicates that the two types of building stones differ only slightly in iron content, but considerably in their deterioration behaviour.

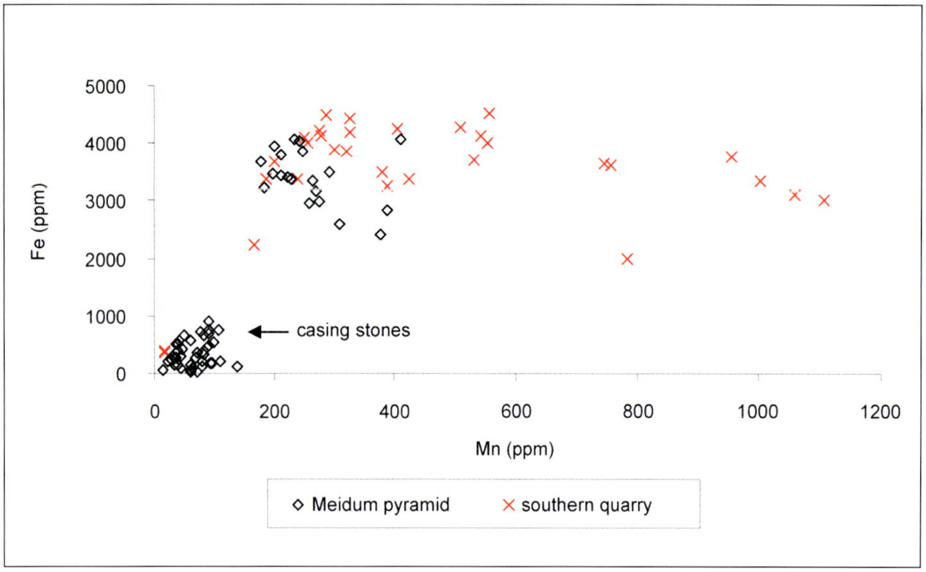

Fig. 42: Fe/Mn-diagram with a clear separation of the low Fe- and Mn-containing casing samples. A rather good correspondence of the core samples with those of the southern quarry is striking. Again, only two quarry samples correspond with casing stones.

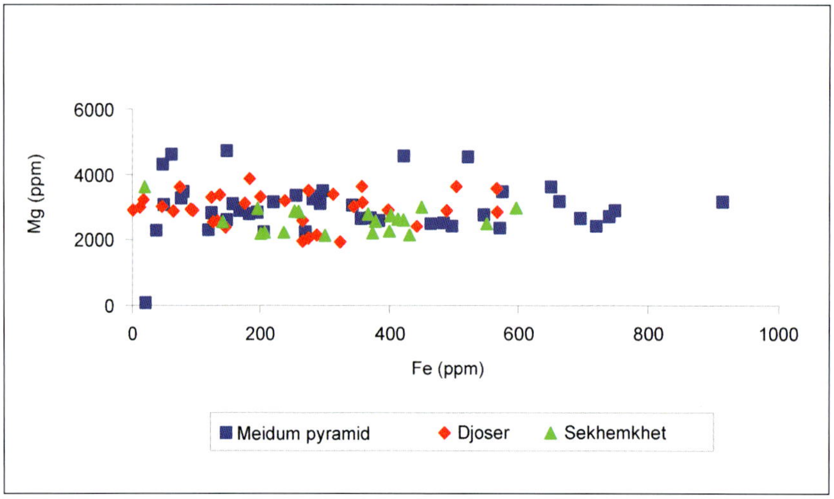

Fig. 43: Mg/Fe-diagram of only the casing stones from Djoser, Sekhemkhet, and Meidum with Fe-values below 1000 ppm. Their almost identical composition is striking.

Finally, in the Fe/Mn-correlation diagram (Fig. 42), the values of the Meidum pyramid's core stones cluster in the field of the southern quarry material, but again more in the part of lower Mn-values. However, the casing material, in spite of nearly the identical petrographic appearance, plots within a separated field with lower Mn- and lower Fe-values. Only two samples from the present day remaining quarry site are close to this chemistry. The only explanation for this discrepancy is that both stone types originally occurred in the quarry area south of the Meidum pyramid and that the highly weather-resistant type that was less abundant was mined off, leaving only a few remnants in the quarry site.

The close relation between the casing stones of the pyramid complexes from Djoser, Sekhemkhet, and Meidum is demonstrated in Fig. 43, where the values of the casing limestones are plotted separately with axis of only 1000 ppm Fe and 6000 ppm Mg. The geochemical similarities of these three pyramids' casing material are obvious, even when a few of the Meidum values are of a somewhat increased Fe-content.

From these results, a new view on the rather remote position of the Meidum pyramid emerges: Due to the exhaustion of the high quality casing stone at the Saqqara plateau, the successor of Sekhemkhet could only find an equivalent quality of stone in the Meidum area. He decided to erect his pyramid near to the stone source, for both economic and pragmatic reasons. From there he used the abundant lower quality for the core construction and the less abundant highly weather-resistant variety for the casing.

Snofru then started the experiment to change the original Meidum step pyramid into a true pyramid but soon realized that the high quality stone material was no longer available due to the source exhaustion. Consequently, he gave up this experiment, began the construction of the two Dahshur pyramids further to the north and looked for a new source of fine limestone.

V. The Pyramids of Snofru (2639–2589 BC) at Dahshur

King Snofru, the founder of the 4th dynasty, built his two stone pyramids about 8 km south of Saqqara on the plateau of the eastern desert rim. The northern one is called the Red Pyramid, whilst the name of the southern monument, about 1.5 km away, is Bent Pyramid. The pyramid is so called due to its striking shape, resulting from an approx. 11° reduction of the slope in the middle of the building.

1. *The Bent Pyramid*

Geographical position (centre of pyramid): 29°47'25" N, 31°12'33" E

Height: 104.71 m | Height of lower part: 47.04 m | Height of upper part: 57.67 m | Base: 189.43 m
Base at bend: 123.58 m | Slope of lower part: 55° | Slope of upper part: 43° | Length of causeway: 704 m

Due to its outer structure, the Bent Pyramid (Fig. 44) is a curiosity among the Egyptian stone pyramids, but it is also the first planned true pyramid. It is the best-preserved pyramid in Egypt as most of its casing remains until today. Thus, the outside appearance of the pyramid gives a most solid impression in comparison to almost all other pyramids in Egypt.

Despite its attractiveness, systematic archaeological investigations only started in modern times with FAKHRY[42].

The original construction plan intended a smaller pyramid with an inclination angle of 58–60°, which would have resulted in a total height of 137.5 m if the plan had been realized. It seems that this plan was followed until a height of about 49 m. Then the angle was reduced to 54.5°, which consequently caused a broadening of the base. The motive for this modification of the construction plan might be a result of cracks and breaches, which appeared in the internal masonry, with the hope that a broadening of the base would increase the building's stability. A further reduction of the angle to only 42° from about 49 m height also reduced the load on the substructure significantly, so that it remains stable even today.

Internal structural damages can be caused by careless masonry construction work. More likely, however, is a sagging of the bedrock, for which a clay-rich substratum can be responsible, but this is rather unusual, as undisturbed sedimentary rocks – even when rich in clay – normally remain stable. More likely is a collapse of the karst systems in the sandy limestone formation of the bedrock. Such karst systems occur in the entire Eocene limestone formation of Egypt. Mainly in sandy limestones and in tectonically affected areas, the bedrock may become brittle and collapse, particularly when heavily burdened. Studying stereoscopic aerial photos or modern high-resolve satellite images,

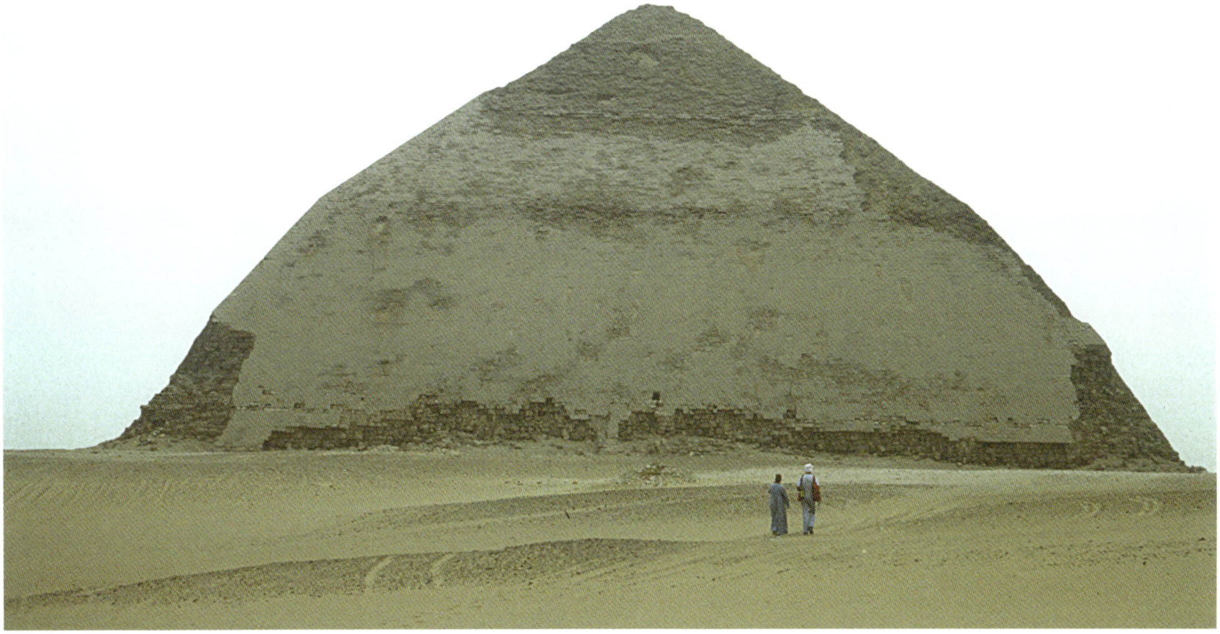

Fig. 44: Bent Pyramid of Snofru in Dahshur, seen from north.
Note the still well-preserved casing, yet mostly destroyed at the corners and the base.

one quickly becomes aware that, especially in the Dahshur area, quite a number of doline-type sinkhole structures can be observed that were most probably caused by karst collapses. However, this assumption of the Bent Pyramid's structural damage still needs geophysical confirmation. Nevertheless, the pyramid was most probably abandoned because of these dangerous cracks and was never used as a burial place[43].

The pyramid complex includes a valley temple, which was built with fine white limestone, and with an entrance in the middle of the south facade at the edge of the escarpment. It is currently under excavation. From its south-west corner, a causeway leads to the north-eastern corner of the pyramid's huge enclosure wall, which in contrast was made of local yellowish-grey limestone.

A cult chapel with fine white limestone walls and roof once stood at the eastern side of the pyramid, directly on the east-west axis. Here, an altar was located and two 9 m high limestone stelae on its north and south sides. At the southern enclosure wall, a small cult pyramid was built in the central axis of the pyramid.

Core masonry

Due to the well-preserved casing and backing stone layers of the Bent Pyramid, a satisfying statistical sampling of the true core stone types is rather difficult and is best possible only at the corners (Fig. 45). A further problem is the search for a nearby quarry pit, as all possible candidates suffer from heavy replenishment of sand and consequently sand blow of outcropping rocks, destroying possible chisel marks.

A petrographic comparison of the core samples with rock specimens of the pyramid's environs gives clear evidence of their close similarity. Thus, the source of the core material must be located in the rock units of the slightly sloping plateau on which the pyramid is situated, most probably in its near vicinity.

The average core masonry, as far as exposed for sampling, is beige to yellowish-brown, sandy limestone.

Macroscopically and with a hand lens, the limestone exhibits intensively calcified fossil remains of mainly small shells and frequently small nummulites, which are generally fragmentised. The sand grains are hard to recognize even with a hand lens but give the broken rock surface a typical sandstone feeling when stroked with a finger. Nevertheless, when a drop of diluted hydrochloric acid is applied, the characteristic bubbles appear, clearly indicating the limestone nature. Especially the more beige, marly limestone types appear (particularly) powdery at weathered parts of the surface due to their loosened clay content.

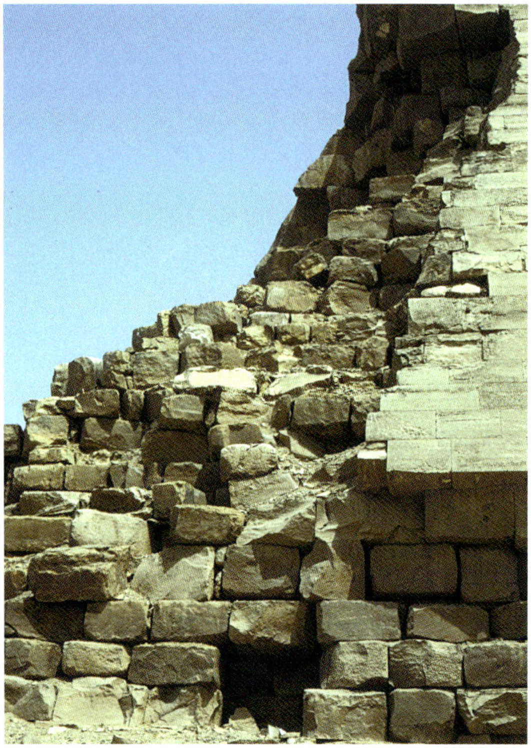

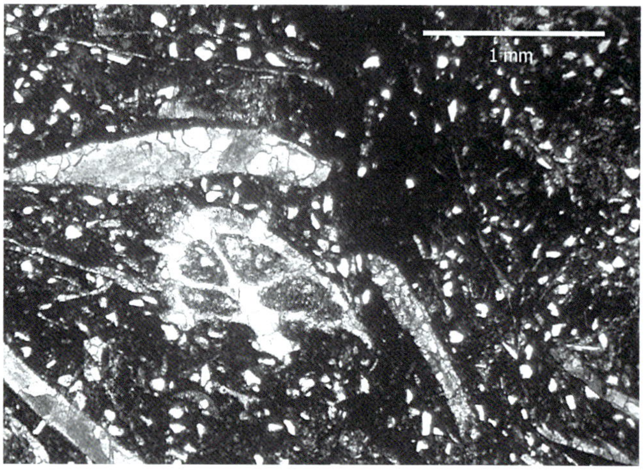

Fig. 46: Micro-photo (Sk 29) of sandy limestone of the core masonry of the Bent Pyramid with intensively calcified micro-fossil remains, almost impossible to identify palaeontologically (1 pol. filter).

◄ *Fig. 45: North-east corner of the Bent Pyramid exposing both core and casing material with backing stones.*

Under microscope, many of the core rock samples were composed of a sandy to very sandy limestone, partly rich in fossil remains, which are hard to identify because of intensive secondary calcification (Fig. 46). The mostly evenly distributed quartz grains vary from only 0.05 to 0.2 mm, and they are always sharp-edged. The matrix is a fine micritic limestone darkened by both extremely fine-grained clay and limonite. The fossil remains consist mainly of shell, small snail, and corral particles.

Casing material

The casing is generally well-preserved, but in some parts it has suffered greatly from deterioration and stone robbery. This is especially the case at the western side (Fig. 47) as well as on the edges, top, and base of all sides. Originally, it was a fine, whitish-grey limestone with a well grinded surface that has been tinted during the many years of exposure to a light brownish colour. This change in colour is due to the migration of oxidized iron containing moisture from the stones interior toward the surface.

Macroscopically, the fresh whitish-grey casing material resembles the fine limestone qualities of Tura and Maasara from the escarpment on the opposite eastern bank of the river Nile. It is of the same fine porous dense composition, but can contain small, calcified shell fragments.

Under microscope, the limestone samples have, however, more similarities with those from Maasara due to their coarser grain sizes and richer content of micro- and nano-fossils, relatively compared with the Tura limestones. The groundmass consists of fine-milled, partly re-crystallized fossil debris, giving an almost sparitic impression. All fossil remains are secondarily calcified, so that a precise palaeontological identification is impossible. Nevertheless, remaining ghost structures of globigerinae, echinoderms, lamellibranches, and sponge needles are observable in each thin section, regardless of their state of preservation.

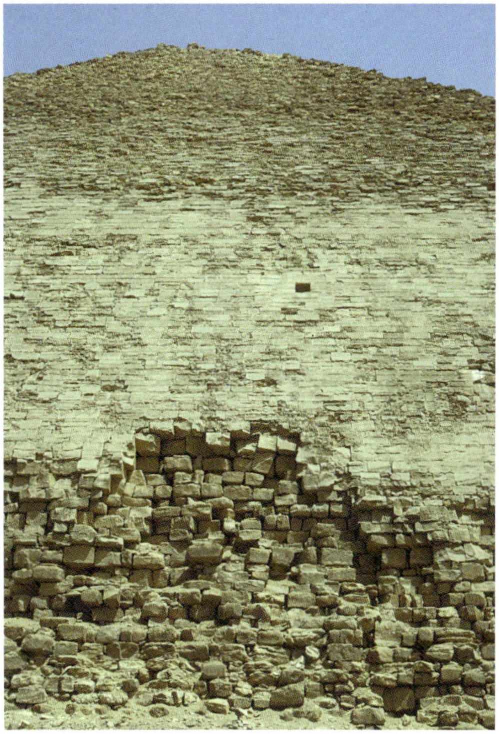

Fig. 47: Heavily weathered casing at the upper part of the Bent Pyramid's western side.

The quarry region of Tura and Maasara

As petrographical and geochemical studies have demonstrated that the casing limestone of all the pyramids from the reign of Snofru onwards generally derives from this quarry area, it is necessary to give a description of it.

Opposite the chain of pyramids on the western bank of the river Nile, this quarry area starts in the Mokattam plateau south-east of Cairo and runs southward until an impressive limestone range forms a steep escarpment that slopes towards the Nile valley. The area is then interrupted by several wadis and finally extends from the desert in a predominantly east-west direction into the Nile valley. Huge opencast quarries, especially along the Mokattam range, that have mainly been used for housing construction during all historic stages of Cairo are increasingly overbuilt by modern industrial enterprises. The dark entrance holes of several gallery quarries used during the New Kingdom can still be seen from the main asphalt road. Following the street to Mokattam City, partly collapsed entrances of older gallery quarries that date to Old Kingdom times are easily recognizable. The whitish-grey limestone, at least in these older quarries, is of fine-grained, dense structure and resembles in its appearance the casing limestones

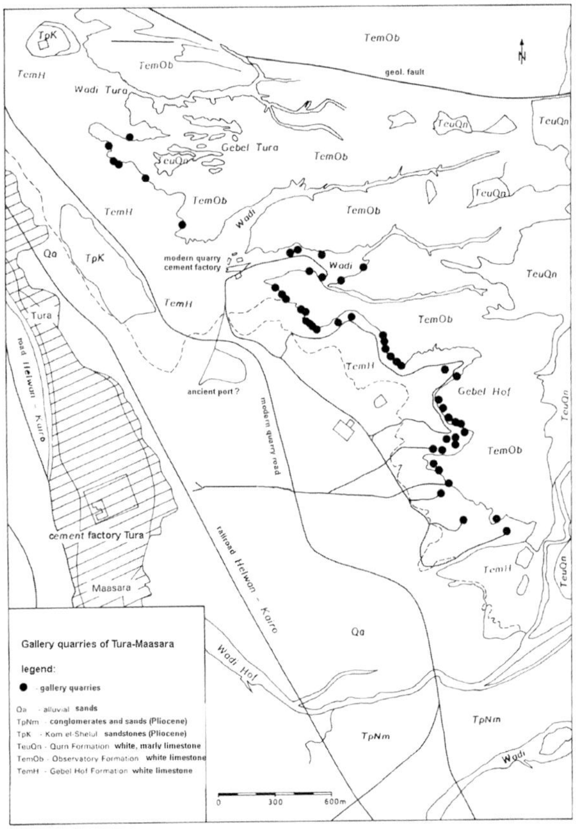

Fig. 48: Topographic map of the quarry area of Tura-Maasara with positions of gallery quarries (black dots).

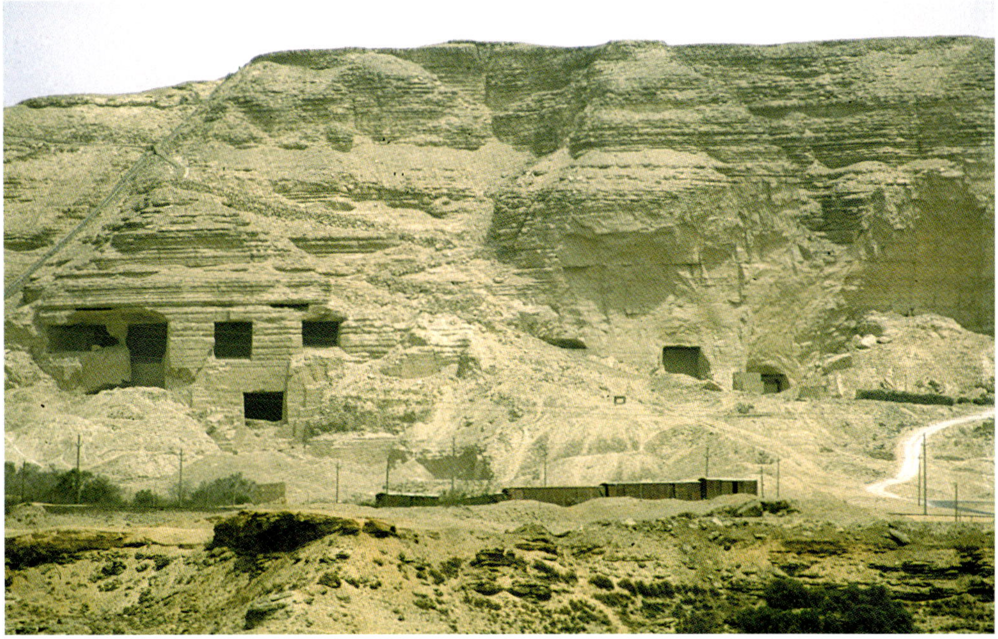

Fig. 49: Entrance holes of gallery quarries at Maasara, view from the road Cairo-Helwan.

used in most of the stone pyramids. However, detailed petrographical investigations have ascertained that the internal microstructure of these Mokattam limestones differs too much to be the general source of the post-Snofru pyramids, even though geochemically some relations are strikingly similar[44].

The next escarpment sector that has, due to the sheer number of gallery and opencast quarries, been worked intensively, roughly extends from Wadi Tura to Wadi Hof (Fig. 48). Within this extended area, numerous gallery quarries can be counted when driving along the Cairo-Helwan asphalt road (Fig. 49). However, with the exception of the site around the Tura cement factory, the entire area is occupied by military camps, which use most of these galleries as army magazines. It is therefore extremely difficult to receive any permission to visit and sample these sites. We were very fortunate to sample the quarries around the cement factory and most of the gallery entrances along Maasara, but the interiors and the foothill opencast quarries could not be investigated due to above-mentioned circumstances. However, the successfully procured samples enable a preliminary study of both the petrography and geochemistry of the entire area.

Despite this somewhat incomplete inspection of the Tura-Maasara quarry area, it could still be established, after studying the chisel marks on the quarry walls, that the individual quarries were used not only during the Old Kingdom for pyramid and temple constructions, but also throughout the entire Pharaonic period including Ptolemaic times.

Petrographically, the rocks from Tura and Maasara are fine to very fine-grained, dense, whitish-grey limestones, which are marly in parts (again, the tongue sticks to the fresh surface).

Macroscopically, the Tura and Maasara limestone types look very similar; only the Maasara limestones are slightly coarser-grained compared with examples from Tura. No fossil remains can be detected.

With a hand lens, the slightly coarser impression of the Maasara type becomes more evident, but it seems to be more compact in comparison with the more fine, porous Tura type. No fossils or fossil remains are recognizable.

Under microscope, again both types are very similar, but the coarser structure of the Maasara limestone, due to very fine-milled fragments of fossils, is noticeable.

The very fine-grained Tura limestone is of a micritic to microsparitic matrix with a characteristic fine porosity. Sponge needles and remains of gastropods and shells are only occasionally visible (Fig. 50).

The somewhat coarser grained Maasara limestone shows, at least at a higher magnification, fragments of fine-milled fossil remains, mainly of sponge- and echinoderm needles, but also occasionally of gastropods, globigerinae etc. (Fig. 51).

Stratigraphically, the fine limestones of Tura and Maasara belong to the Mokattam group and here into the Wadi Hof formation, which corresponds internationally to Lutetian.

Geochemical provenance determination of the core material

The analysed data, including the small cult pyramid, are shown in the correlation diagrams of Figs. 52–54. In all three diagrams the plotted data of the Bent Pyramid core masonry cover an unusually wide field. This dispersion does not really fit with the distribution field of the rock samples that were taken from the various outcrops around the pyramid. This result contradicts with the petrographical observations, which show a rather good concordance of the core masonry with the rock fabric of the outcrops surrounding the pyramid. Further, the relatively high Fe-values in comparison with the Red Pyramid core material (Figs. 64–66) are somewhat remarkable, as the reddish colour of the latter derives from iron oxide, and therefore greater iron content was expected.

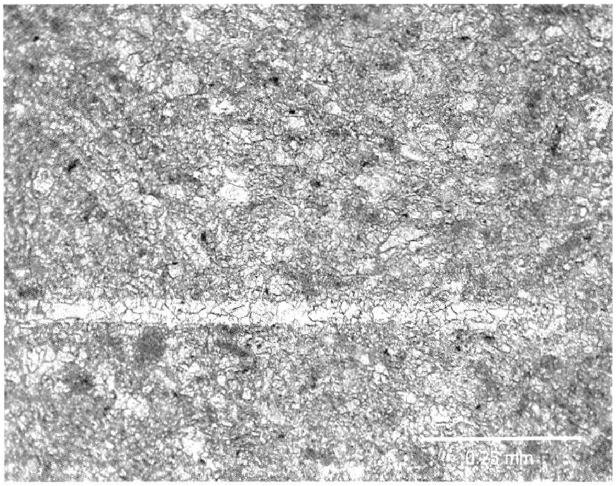

Fig. 50: Micro-photo (QS 229) of Tura limestone. Note the calcified sponge needle and in comparison with Fig. 51 the lower sizes of the micritic calcite grains (1 pol. filter).

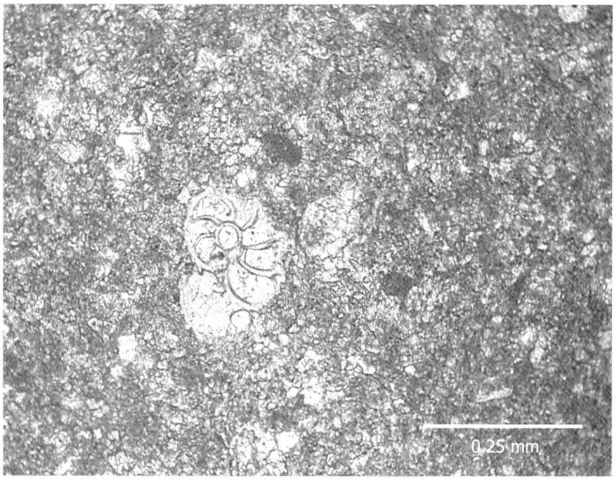

Fig. 51: Micro-photo (QS 1160) of Maasara limestone. The slightly coarser structure of the calcite grains and the (even rarer) globigerines is the main difference to Tura (1 pol. filter).

The only explanation for these discrepancies is, in combination with the petrographical observations, that most of the core material derived from yet unidentified quarry sites, most probably due to the considerable sand coverage of the entire area.

Yet, after studying stereoscopic aerial photos in combination with high-resolution satellite images of the area around the Bent Pyramid, it can be assumed that the source of the pyramid's core masonry might be to the east of it, in a heavily sand-filled depression, which would be hard to define as a quarry in the field. It shows artificially looking steep rock structures at its southern and eastern rims. This site is all the more probable, as there is a ramp leading directly up to the pyramid's eastern side (Fig. 55).

Unfortunately, during a later sampling campaign, both sampling and export of samples for scientific analysis were prohibited by the antiquities authority, so that we could not prove this assumption.

Geochemical provenance determination of the casing material

All geochemical correlation diagrams of the casing samples (Figs. 56–58) show a satisfying correspondence of their data sets to those of the Maasara quarry area, which, however, we could not subdivide in more detail, as most of the many galleries are under military control and therefore impossible to sample. Comparing the distribution fields of the Bent Pyramid casing samples with those of the Maasara geochemical data, it becomes apparent that a considerable amount of the Bent Pyramid data is densely plotted in a small field, indicating that the pyramid samples might occur from a very limited part of the extensive quarry area. Likewise, the petrographical observations also point to Maasara quarries, supporting this provenance determination. Due to the closer position of this quarry area to Dahshur, STADELMANN[45] also suspected that the Maasara sites were the source of the Bent Pyramid casing material. This assumption can now be confirmed scientifically.

Stratigraphically, the geological units supplying the core masonry of the Bent Pyramid belong, according to the geological map of KLITZSCH/LIST/PÖHLMANN [46], to the Kom el-Shellul formation of the Pliocene (Upper Tertiary), which is described as shallow marine sediments of calcareous sandstone and coquina limestone[47]. However, according to the Geological Map of the Greater Cairo Area[48], the units date to the Pliocene (undifferentiated) but apparently consist of non-marine conglomeratic sands and sandstones and gravels up to 18 m thick. This description does not really fit to the above described rock types, which form the core masonry of the Bent Pyramid.

During investigation of the area directly surrounding the two Dahshur pyramids, non-marine fluviatile sediments with a maximum thickness of 1 m and partly solidified to conglomerates could be detected in certain places. But next to the pyramid, marine sediment crops out, which shows, at least in its lower portions, a conspicuous similarity to the rock units occurring around and at the bases of the pyramids of Pepi I and II, Merenre, and Djedkare-Isesi in Saqqara-South. However, the geological environment of these pyramids belongs, according to the Geological Map of Greater Cairo[49], to the Middle Eocene Ravine Beds of marl and marly limestone with clay interbeds. The strata delivering the core material for the Bent Pyramid have some relations to the Saqqara Member of the Late Eocene Maadi formation but consist rather of sandy and marly limestone with interbeds of shale. To the west of the Dahshur pyramids, these units are overlain by the Pliocene Kom el-Shellul formation. Thus, according to YOUSSEF ET AL.[50], they could also be identical with the Giran el-Ful Member of the Maadi formation.

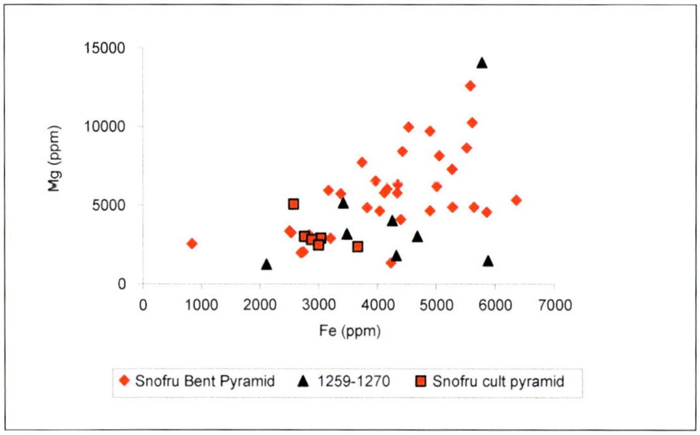

Fig. 52: Mg/Fe-diagram of Bent Pyramid core limestone in comparison with nearby limestone outcrops. Note that most of the outcrop samples plot within or next to the field of the core stones, but without a satisfying concordance. This and the petrographic concordance of pyramid and assumed quarry samples indicate that the true pyramid quarry must be in the near vicinity.

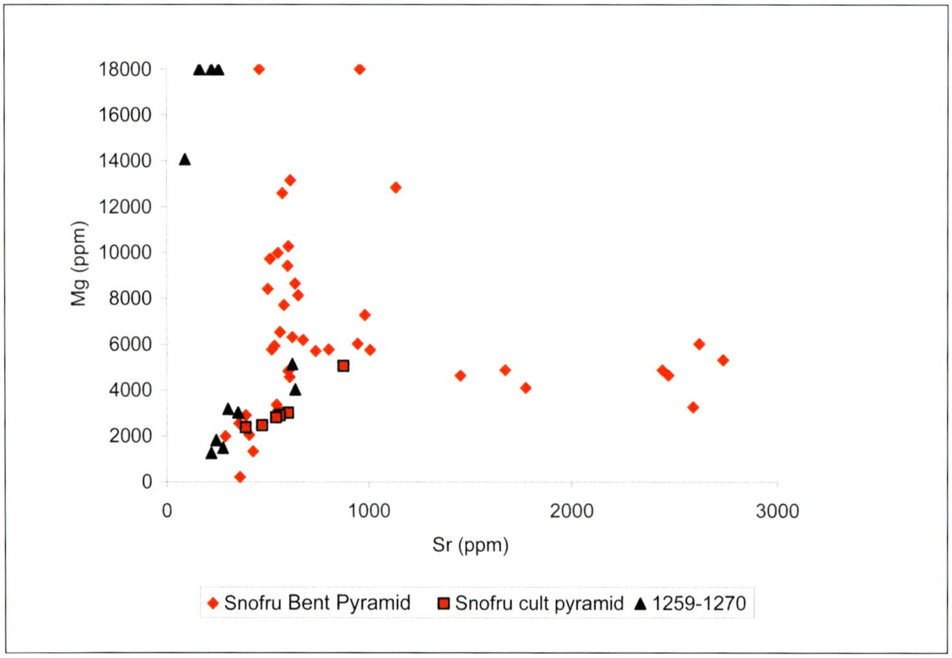

Fig. 53: Mg/Sr-diagram of Bent Pyramid core limestone in comparison with nearby limestone outcrops. A rather poor coincidence between pyramid and sampled stones of the pyramid's vicinity indicates that the true quarry site was not sampled.

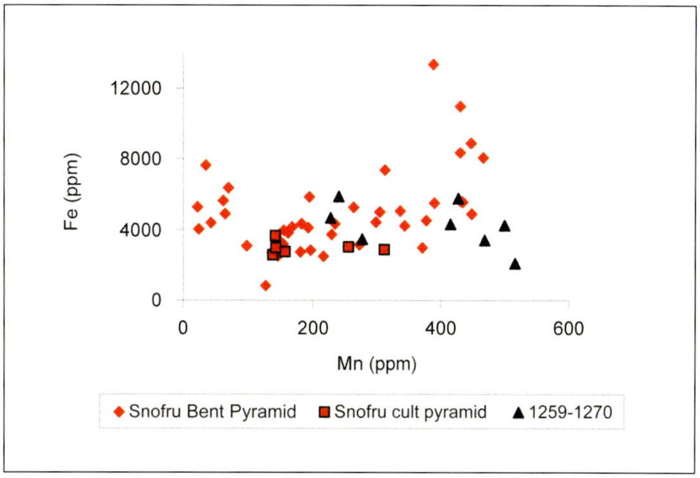

Fig. 54: Fe/Mn-diagram of Bent Pyramid core limestone in comparison with nearby limestone outcrops. It shows the same result like Figs. 52–53.

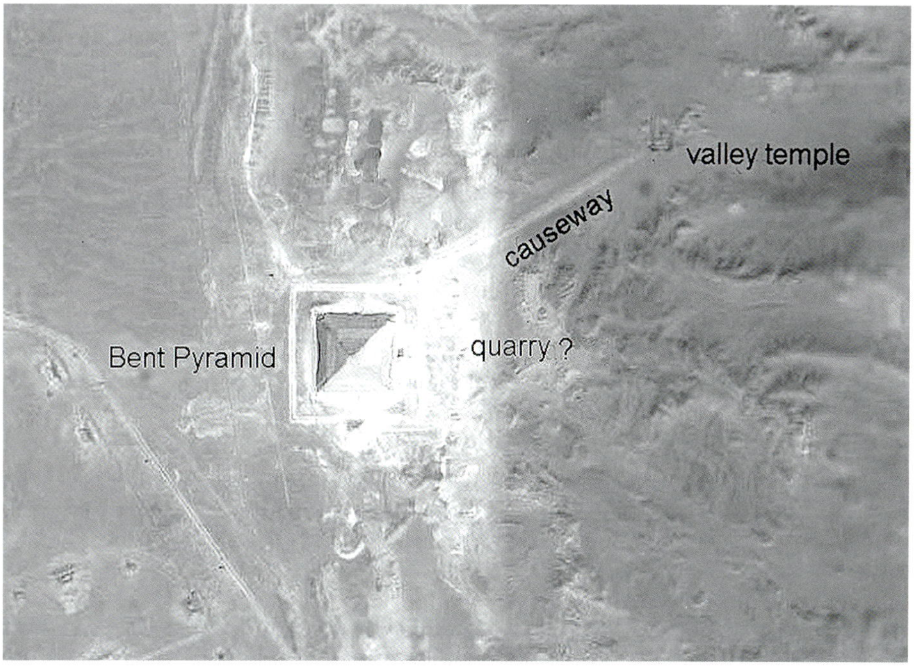

Fig. 55: Bent Pyramid of Snofru with possible main quarry site (modified Google image 2006).

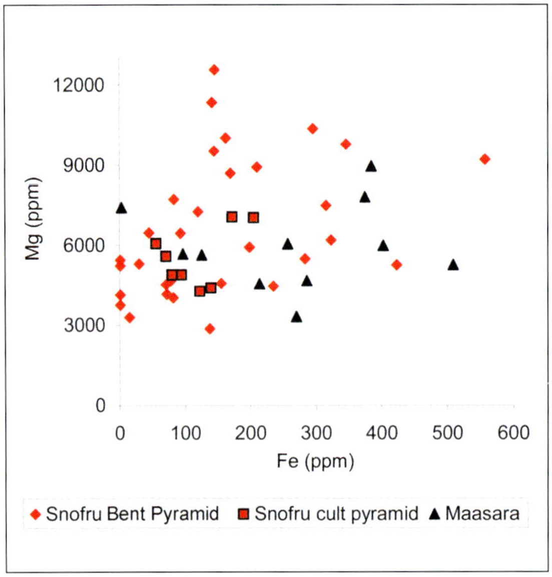

Fig. 56: Mg/Fe-diagram of Bent Pyramid casing limestones. This diagram allows a rather good, but not really a close attribution of the Bent Pyramid casing material to Maasara. However, the petrographical observations also point more to Maasara.

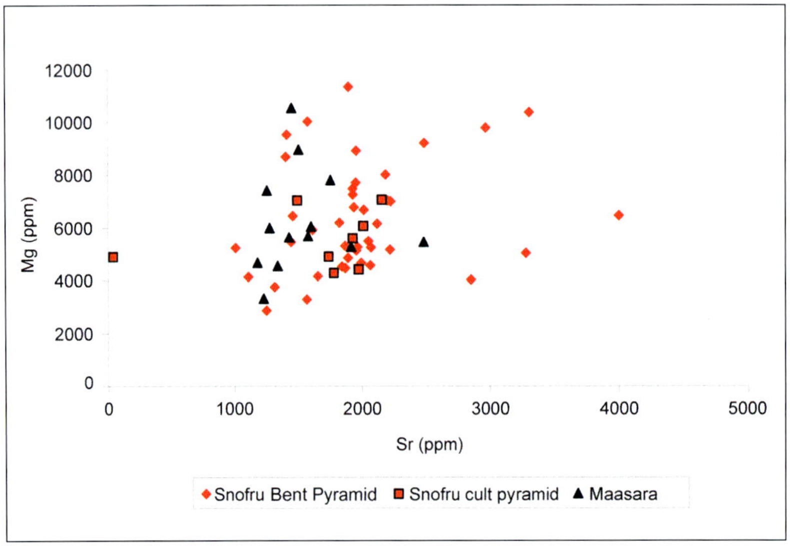

Fig. 57: Mg/Sr-diagram of Bent Pyramid casing limestones. Without a slight tendency to higher Sr-values, the majority of the casing stones plot into the field of Maasara.

Fig. 58: Fe/Mn-diagram of the Bent Pyramid casing limestones. Poor in both iron and manganese, these limestones plot namely close to Maasara but also slightly apart. This may indicate that the true Snofru quarry is not precisely identified.

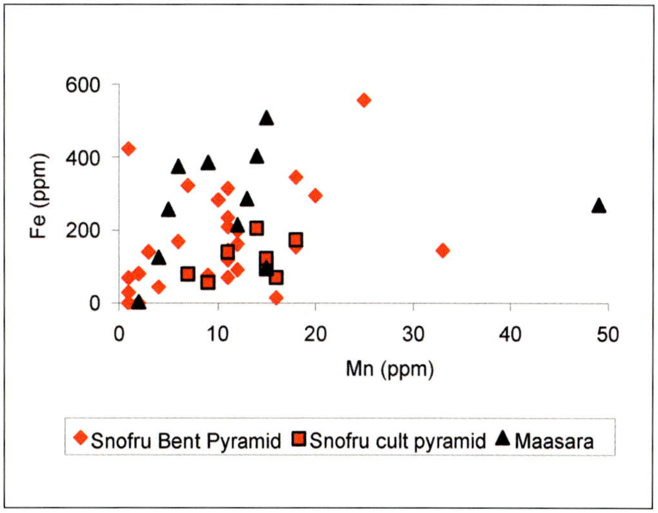

2. *The Red Pyramid*

Geographical position (centre of pyramid): 29°48'31" N, 31°12'21" E

Height: 104 m | Base: 220 m | Slope: 45°

Red Pyramid quarries: 29°48'11" N, 31°11'50" E
29°48'21" N, 31°11'58" E

About 1.5 km north of the Bent Pyramid stands the so-called Red Pyramid (Fig. 59), which is also attributed to Snofru. The pyramid's casing limestone has been almost completely dismantled revealing its bare core. The core consists of a reddish-brown, sandy limestone, which gives the pyramid its familiar name.

Archaeological investigation of the pyramid first took place in the first half of the 19th century, but a systematic and detailed excavation was carried out only in the last decades by STADELMANN[51].

The Red Pyramid was probably built after the Bent Pyramid. In the case of the Red Pyramid, it is clear that construction risks were carefully avoided in contrast to the previous monument. Thus, the foundation was laid upon a more solid ground with three layers of large, fine limestone blocks, which were extracted from the Tura-Maasara quarries (see below). The slope was reduced to 45°, causing an enlarged basis compared with the Bent Pyramid, which consequently reduced the pressure on the limestone surface significantly.

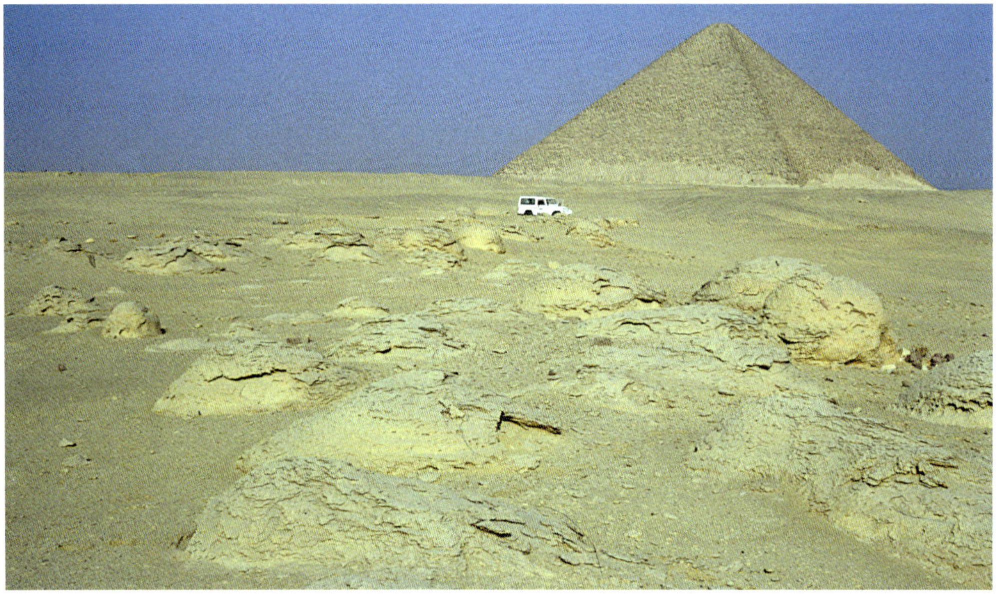

Fig. 59: Red Pyramid of Snofru at Dahshur, seen from its south-western quarry site.

It is the fourth highest pyramid ever built in Egypt, with almost 160 layers of stone. Further, the Red Pyramid was the first successfully built true pyramid in Egypt, a style that was repeated in the later Gizeh pyramids.

Some of the remaining casing blocks show the cartouche of Snofru and marks of labourers. These inscriptions present to us information not only about how long the pyramid took to be built but also about the sequence of work that took place. From these markings, we know that the pyramid was probably begun between year 22 and 29 of Snofru's reign, and that the entire pyramid was finished in approx. 17 years.

To the east of the pyramid stand the remains of a mortuary temple. In this area, the first capstone (pyramidion) ever found belonging to an Old Kingdom pyramid was discovered. It was recovered in fragments and reconstructed. The mortuary temple itself, consisting of nothing more than remains, is significant, because Snofru pioneered the east-west alignment of Egyptian temples. As in most Egyptian pyramids, the entrance lies at the northern side.

Core material

The source of the Red Pyramid's core material is not at all enigmatic, because two (still) clearly recognizable transport ramps run from the pyramid's south-west corner directly into an extensive quarry area (Fig. 60). A small number of heavily weathered stone blocks remain there that are of the same size and sandy, reddish-brown appearance as used in the core masonry (Fig. 61). The outcropping rock units appear to be somewhat brittle, but fresh surfaces show a hard consistency, even though the stone is partly highly and irregularly porous. Unfortunately, parts of the ancient quarry site are off limits, closed off by an army fence and therefore not accessible for sampling.

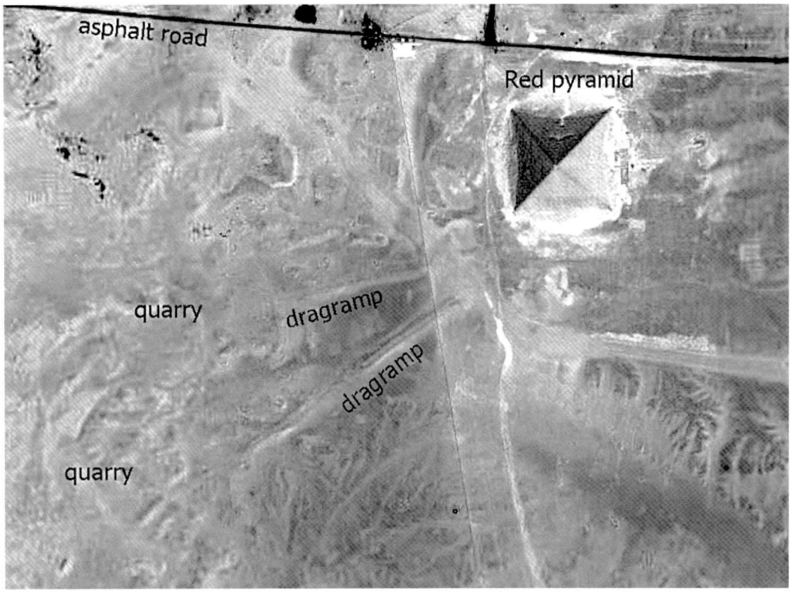

Fig. 60: Overview of Red Pyramid area (modified Google image 2006) with two drag ramps running into the quarry sites.

Even under the dry Egyptian weather conditions, these sandy limestones to almost calcareous sandstones become, when exposed, surprisingly unstable and increase their reddish-brown tint. Their fossil content of mainly secondarily calcified shells and oysters quickly starts to disintegrate, leaving negative forms of their original structures, thus destabilizing the former solidity of the stones even further.

As the Red Pyramid suffered intensive dismantling of its original casing from early Arab (?) times, the core stones were open to the elements and were affected more by weathering than most of the other monuments situated further to the north, which used more robust building stones.

Macroscopically, the appearance of the various building stones depends very much on the exposure conditions. If the original material contains large fossils, such as oysters, that gradually decrease the homogeneous state, then it is more susceptible to the effects of weathering agents such as sand and water. In general, the building stones of the Red Pyramid are of a sandy character, and it is difficult to differentiate between sandy limestone and calcareous sandstone when attempting to classify them. The colours vary between yellowish-beige and reddish-brown, depending mainly on the degree of exposure. Many types of fossils form a rich population, from which palm of the hand large oyster shells are the most striking. Apart from these, remains of corals and many types of smaller shells are macroscopically recognizable. The latter are always secondarily calcified and thus hard to classify. Nummulites could not be detected.

With a hand lens, the sandy consistency comprising of clear sand grains, especially in the slightly altered specimens, dominates the structure. However, the appearance of bubbles when a drop of diluted hydrochloric acid is applied to the surface indicates the presence of calcite.

Under microscope, at the first glance, the sharp-edged and well sorted sand grains (0.05–0.1 mm), mainly of quartz but rarely of plagioclase, stand out. The micritic carbonate interspace exceeds the sand fraction by far. The

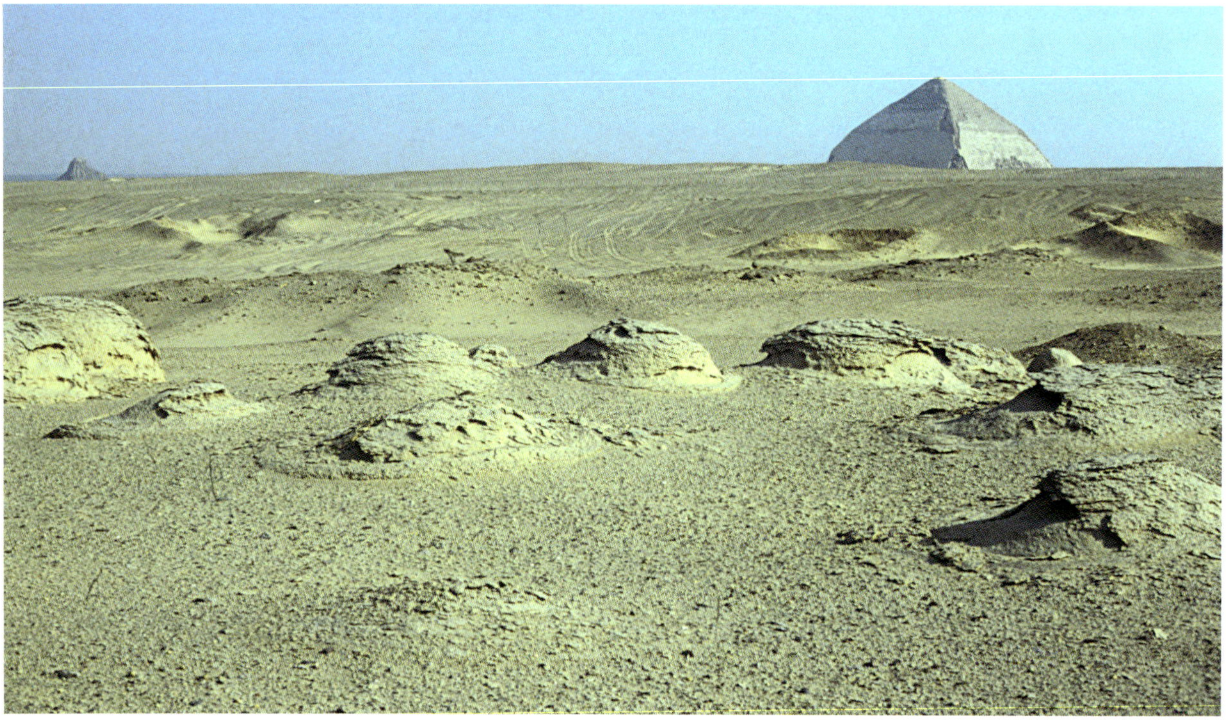

Fig. 61: Remaining highly weathered block pattern in the Red Pyramid quarry site, south-west of the monument (Bent Pyramid in the background).

characteristic fossil remains, with the exclusion of the oyster shells, show an intensive sparitic calcification and are therefore impossible to classify (Fig. 62). It is uncertain why only the oyster shells and their fractions remained in their original structure without any secondary calcification.

Geochemical provenance determination of the core material

The microscopical identification of the core masonry's sandy limestones as stone material from the nearby quarry site (Fig. 63), which is still connected to the pyramid by clearly recognizable transport ramps, makes an attribution of the core masonry to this quarry site unquestionable.

However, the geochemical correlation between the pyramid samples and the collected quarry samples is quite satisfactory (Figs. 64–66). Most probably, the inhomogeneous character of the sandy limestones with their variable fossil content and especially the large oyster shells are responsible for their chemical variability. Yet again, samples could not be taken from the whole quarry area due to the military facility.

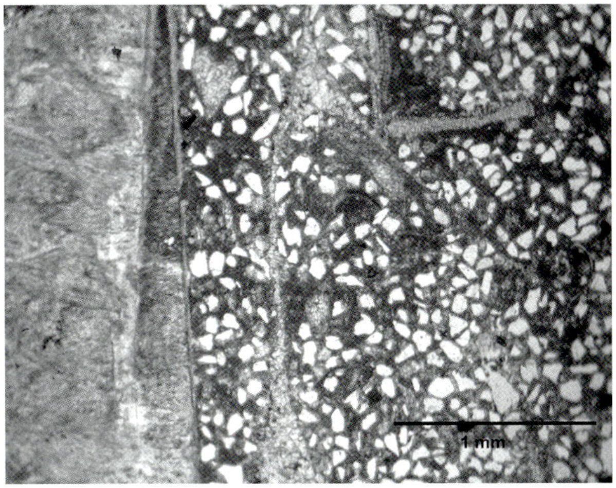

Fig. 62: Micro-photo (Sr 20) of core limestone from the Red Pyramid. The light coloured quartz sand grains are imbedded in a microsparitic dense matrix with completely calcified remains of mainly shell fossils. The oyster shell (left rim), however, is not affected by calcification (1 pol. filter).

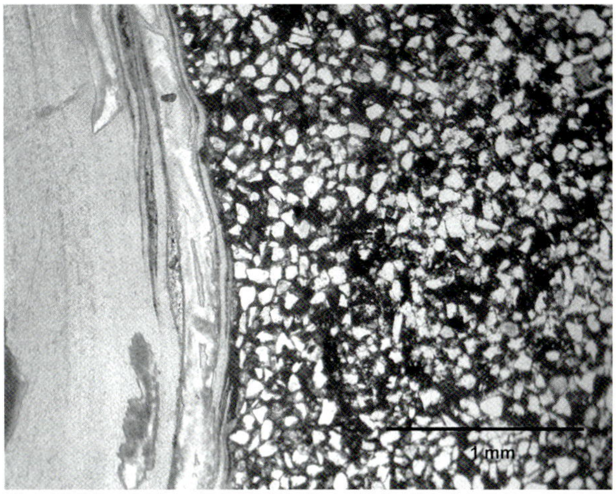

Fig. 63: Micro-photo (QS 1652) of Red Pyramid quarry sample for comparison with Fig. 61. Note that both images are almost identical. But one must consider that the quarry material, as well as the core blocks, have a wide variety of appearance from fossil rich to a dense limonite rich calcareous sandstone (1 pol. filter).

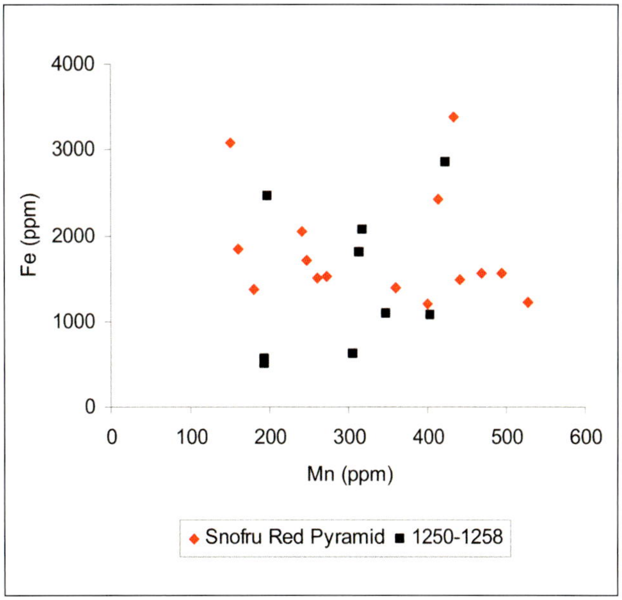

Fig. 64: Fe/Mn-diagram of core limestone from Red Pyramid and nearby quarry material. The relatively low number of quarry samples does not allow a satisfying correspondence.

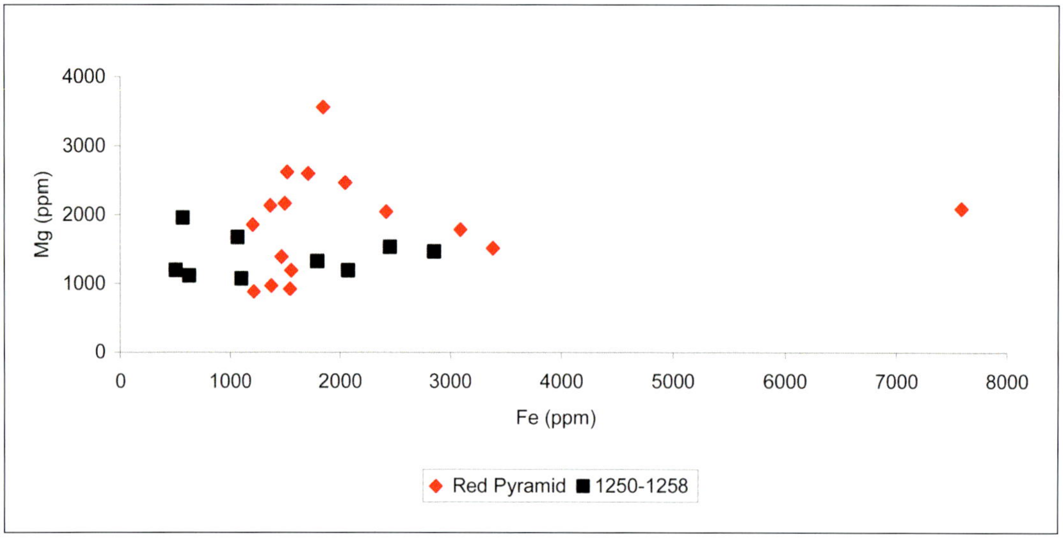

Fig. 65: Mg/Fe-diagram of core limestone from Red Pyramid and nearby quarry material. In spite of the rather poor concordance between pyramid and quarry samples, the petrographic correspondence is unequivocal.

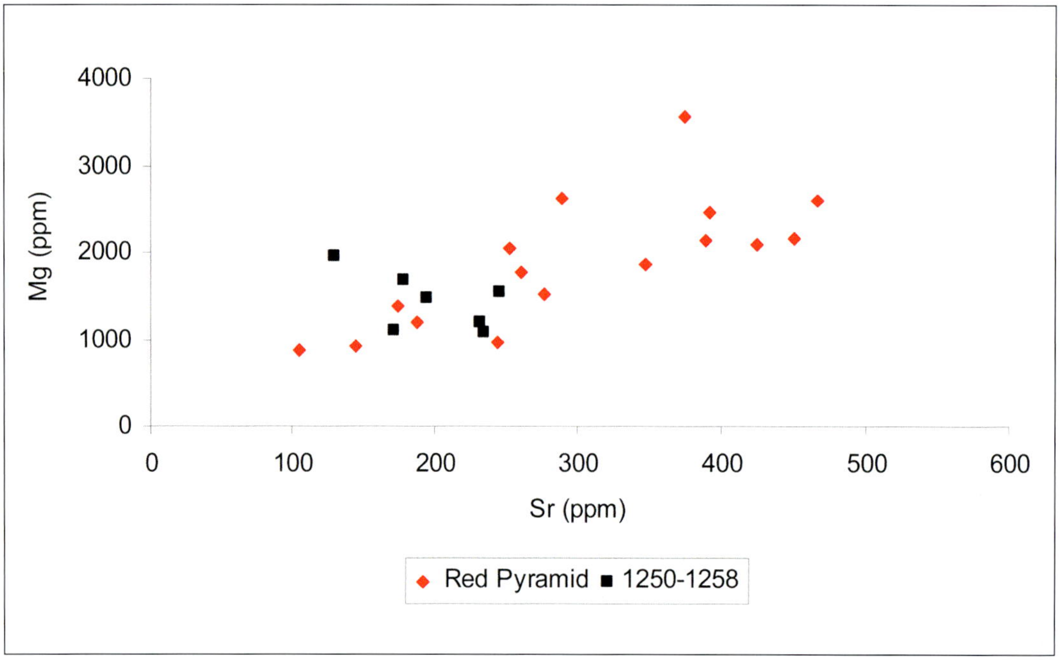

Fig. 66: Mg/Sr-diagram of core limestone from Red Pyramid and nearby quarry material.

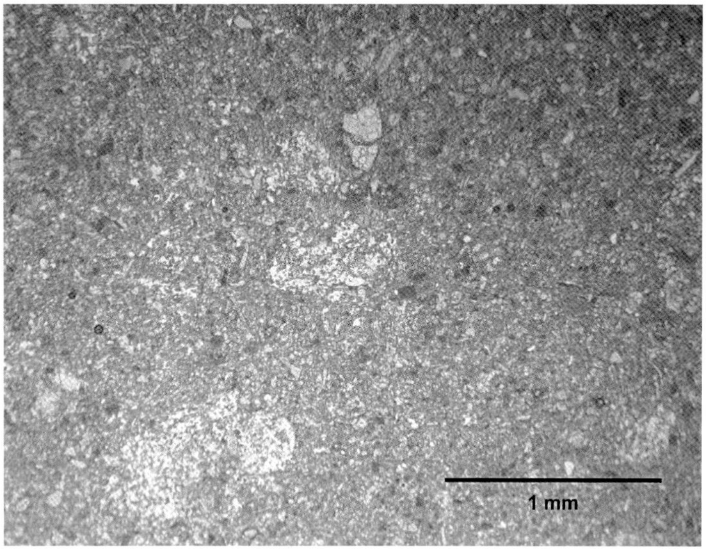

Fig. 67: Micro-photo (Sr 1) of casing limestone from the Red Pyramid. The sparse micro-fossil content is secondarily calcified in a microsparitic matrix (1 pol. filter).

Casing material

In spite of the dismantling of the pyramid's outer casing, enough material remained in the debris around the pyramid and even some blocks were left *in situ*. At first glance, this fine whitish-grey casing material greatly resembles the Tura-Maasara limestones. This provenance could be ascertained by petrographical and geochemical analysis, but there are still some remarkable variations in both cases.

Macroscopically, this limestone is of a coarser-grained structure and the grinded surface appears more compact and less porous than the stones used later in the Gizeh pyramids. Further, the colour tends to be more whitish and never shows the slight yellowish tint observable in many examples from Tura.

With a hand lens, this impression of a compact and extremely low porous stone type is confirmed. No fossils or fossil remains could be detected.

Under microscope, the diameters of the coarser grains point more to the Maasara quarry area than to Tura. The rock fabric is composed of a fine-milled detritus consisting of calcareous fossil remains, cemented with micro-sparite (Fig. 67). The identifiable fossil remains belong to small shells, gastropods, and sponge needles, all of them completely calcified, thus making a palaeontological classification impossible.

Geochemical provenance determination of the casing material

Geochemically, the casing samples clearly point to a Tura-Maasara provenance. However, the very low Fe-values stand out (Fig. 68). This is also the reason for the more whitish appearance of the fresh limestone surfaces and for

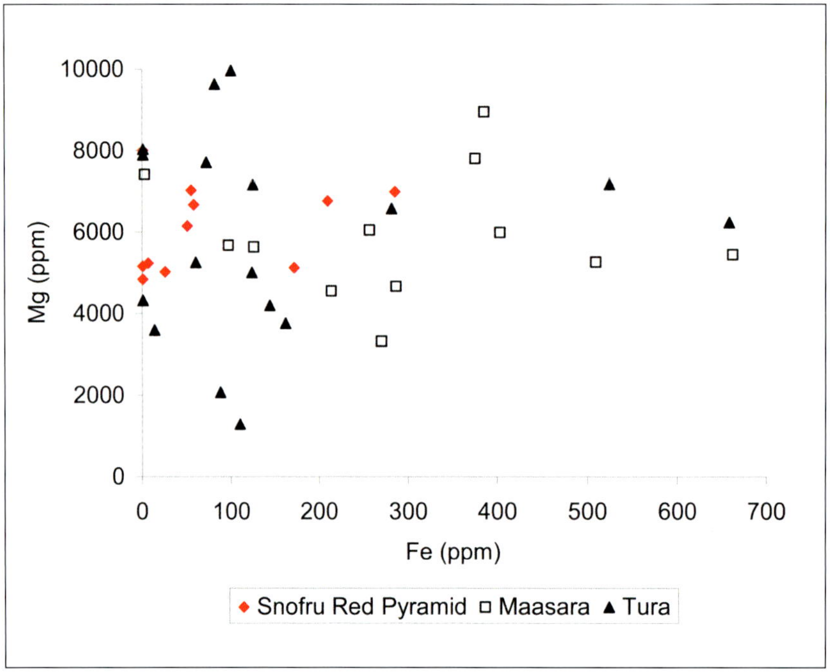

Fig. 68: Mg/Fe-diagram of casing limestone from Red Pyramid. The very low Fe-values plot well into the Tura-Maasara field and here into the very low Fe-part.

the only slightly light-yellowish patina produced in more than four millennia (Fig. 69). Compared with the data sets of casing stones with a classical Tura-Maasara provenance, the values of the Red Pyramid's outer casing appear in a similar distribution field. On the other hand, Sr-values are higher than in the Tura rocks (Fig. 70).

Concerning the Maasara gallery quarries, samples could only be taken at the entrances, as permission was not granted to enter the galleries. However, the limestone at the entrance corresponds, of course, with the stones that line the interior. Nor could samples be obtained from the opencast quarries situated at the base of the escarpment, which were undoubtedly worked during ancient times. As the grain sizes of the Tura-Maasara limestones seem to increase downwards in stratigraphy, we suspect that the Snofru quarries might be located rather within these foothill parts.

Summarizing the data sets of the casing material at Dahshur and comparing them geochemically and petrographically with those of the earlier constructions from Djoser until Huni, a clear change in material provenance becomes evident. The earlier pyramids procured their casing material exclusively from Western Desert sources, i.e. from the Saqqara and/or Meidum area.

Only for the true pyramid E 3-phase of Meidum a number of blocks were extracted from the Tura-Maasara area and conversely, in the casing of the Bent Pyramid, only a few blocks derive from a Western Desert source. This demonstrates that first under the reign of Snofru the provenance of the outer casing stones and fine limestone material in general shifted from Western Desert sources to the Eastern Desert regions of Tura-Maasara.

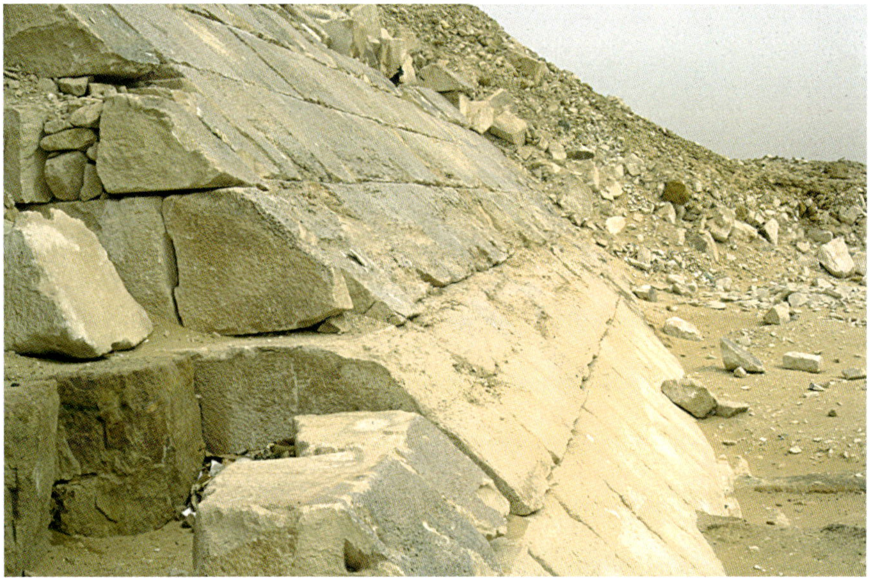

Fig. 69: Part of the white casing of the Red Pyramid, appearing only slightly yellowish after exposure of many millennia.

Regarding the various pyramids built under Snofru, it seems that during this period the development process of pyramid construction reached a temporary solution after starting to transform the originally designed step pyramid in Meidum into a true pyramid. The realization of this new conception of a true pyramid was planned in Dahshur-South, experimenting with different inclination angles. It remains unclear, whether this change came as a result of inside structural damage, as mentioned above, or was planned from the beginning. The presence of cracks, caused by subsidence, supports the first assumption. The next attempt with the Red Pyramid in the northern Dahshur area started with a reinforced and enlarged base (220 m compared with 189.4 m at the Bent Pyramid) and a slope of only 45°.

For pragmatical reasons local sources of stone were used for the core construction, thus it is evident that the availability of raw materials determined the monument's position. In addition, easy access to sufficiently weather-resistant casing material within short distance of the building site also influenced the choice of the pyramid's location, at least until the step pyramid stage in Meidum. Concerning this last aspect, Snofru acted against tradition by obtaining his casing material from Maasara, on the opposite bank of the river Nile, some 25 km away.

In the case of the Saqqara plateau, the position of the first two pyramids was determined not only by the access to sufficient and easily extractable building stone, but also by the proximity to the capital. Comparable arguments for the selection of the Meidum site are less convincing, even when a suitable quarry site for the pyramid's core and casing material is not far away.

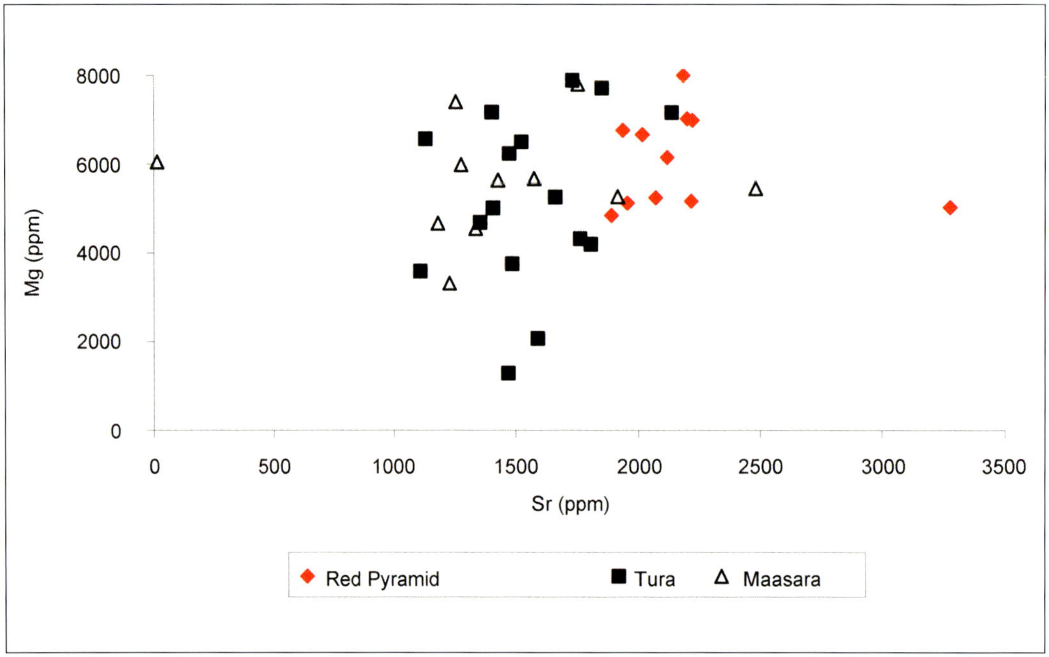

Fig. 70: Mg/Sr-diagram of casing limestone from Red Pyramid. The samples plot into the higher Sr-part of the Maasara field.

The selection of the Dahshur sites is, in view of these aspects, somewhat enigmatic, at least in the case of the southern pyramid. The material available in the immediate vicinity was not easy to extract, as an opencast quarry had to be cut into the plateau. The instability of the basal rock formation was not recognizable to the construction engineers at that time and they cannot be blamed for it. The core material of the northern pyramid could again be procured close to the building site but was of lower quality. Further convincing arguments for the selection of these sites are hard to imagine and will not be discussed here on the basis of mere speculations.

After these experimental phases of pyramid construction, the "ideal" pyramid with an inclination angle of around 52° was developed during the reign of Khufu and served as a model for further generations.

VI. The Gizeh Pyramids

The pyramids of Khufu, Khafre and Menkaure on the Gizeh plateau (Fig. 71) are listed here together due to their close vicinity and mutual geology of the quarry sites, for both core and casing material. The pyramid of Djedefre at Abu Roash is described separately, even when this does not correspond to the general chronological line followed so far.

Geological environment

In contrast to the lack of detailed geological investigation of the pyramid environments at Dahshur, Meidum and to some extent at Saqqara, numerous but unfortunately also contradictory publications exist for the Gizeh plateau. Of these, only the more important will be mentioned in order to offer the reader selected information.

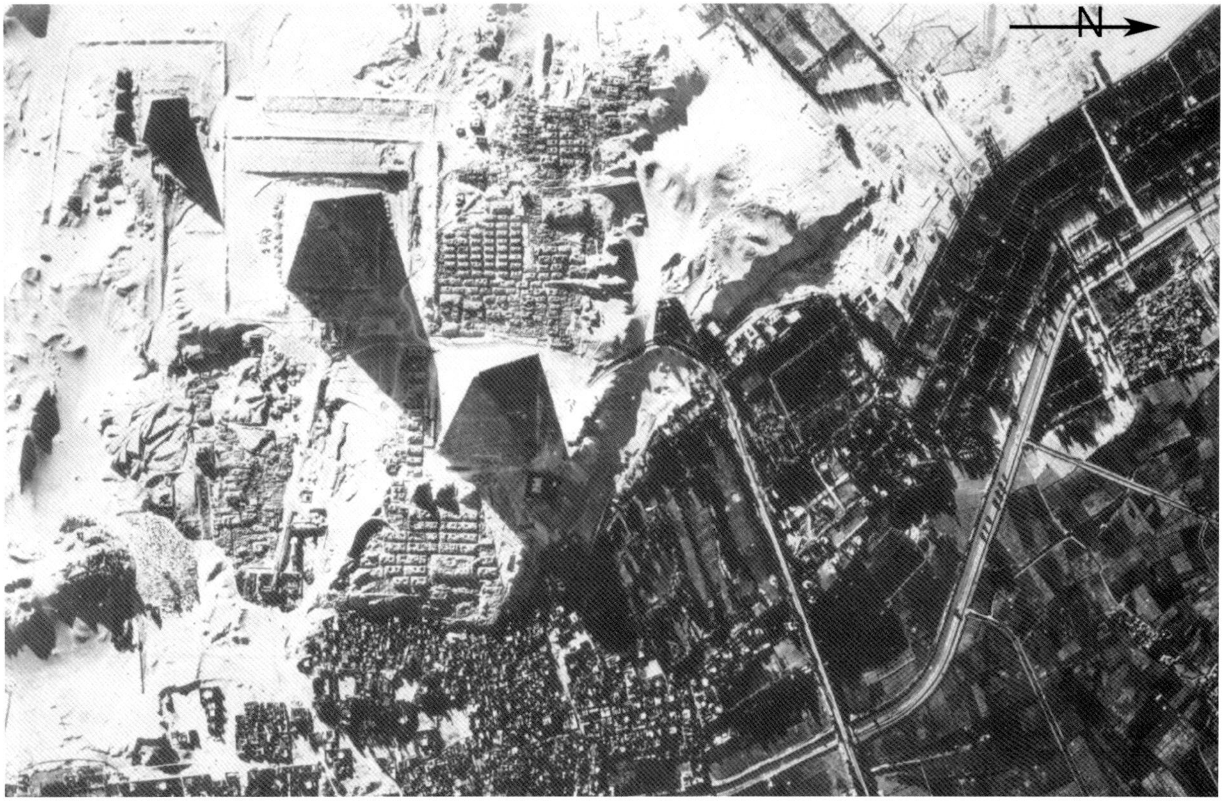

Fig. 71: Aerial photograph of the Gizeh plateau with the Great Pyramids, their archaeological environments and main quarry sites next to the monuments.

Detailed references will, however, be made to the more recent papers of a research group of the Ain Shams University, Cairo and of the American Research Centre in Egypt (ARCE), as they deal with questions directly concerning the provenance of the pyramids' building stones.

First a short historical overview: The first systematic attempt to stratigraphically subdivide the Gizeh plateau was made by VON ZITTEL, who was mainly concerned with the classification of the Eocene fossil groups occurring there[52]. In his handbook on the regional geology of Egypt, BLANKENHORN[53] dealt intensively with the geology of the Gizeh plateau. CUVILLIER revised the Egyptian nummulites and proposed new stratigraphic divisions[54]. HUME, in contrast, did not extensively discuss the Gizeh plateau in his first volume of the "Geology of Egypt"[55]. Interesting geological observations were given by WAGNER[56]. KNETSCH also contributed to the geological situation around the Gizeh area[57]. A detailed stratigraphic division was given by SAID[58], which he later modified[59].

Detailed contributions to the geology of the Gizeh plateau were given by STROUGO together with the group from the Ain Shams University[60]. They presented a well-differentiated stratigraphic division, which YEHIA[61] correlated convincingly with the Eocene limestone sequences of the Eastern Desert south of Cairo. Concerning the building stones, the tectonic details and the special geological map given by YEHIA are of special interest. In his map, geological details and specifications are marked, which must have been apparent to the builders of the pyramids, as not only the exact layout of the pyramids but also the arrangement of the quarry sites were carefully chosen, integrating these aspects into the architectural conception.

Apart from this group of modern Egyptian geologists, international geoscientists have also presented excellent geological contributions concerning the Gizeh plateau. Here, the work of AIGNER[62], which partly deals with topics of geoarchaeology, of GAURI[63], who was also involved in the conservation of the Sphinx, and of LEHNER[64] from the American Research Centre in Egypt (ARCE) should be mentioned with great respect. Especially LEHNER exposed interesting connections between the local geology and the selection of the pyramid positions, the quarry sites, the layout of the entire necropolis, and the port installations. In the case of Khufu's pyramid, he portrayed the development of the whole construction process on the Gizeh plateau from its earliest occupation onwards[65]. Not all of his conclusions can be confirmed in the present monograph. Nevertheless, his careful consideration of the geological and archaeological evidence for the ancient constructions is impressive.

Both LEHNER and the present authors regard the limestone formations that dominate the Gizeh region as Middle to Upper Eocene and the actual limestone of the Gizeh plateau at the pyramid bases as part of the Middle Eocene Mokattam formation.

Concerning the geological classification of the main quarry areas south of the Khufu, east of the Khafre and south-east of the Menkaure pyramid (see below), the two research groups differ somewhat. The ARCE group categorizes the rock units of these quarries into the Mokattam formation (not to be confused with the Mokattam Group), whereas the Ain Shams group classifies them as part of the Observatory formation. In both cases, the Middle Eocene period is accepted. The only discrepancy lies in a facies difference. This "controversy" could only be regarded as an academic divergence, if it did not concern the building materials of the great pyramids and their provenance.

Originally, the stratigraphic classification goes back to AIGNER[66], who understood the hard nummulitic limestones that form the pyramid plateau as a special shallow marine facies, precipitated upon an up-doming zone of the Cretaceous series below. Within a protected area in the flank of this submarine dome, coral reefs could develop; they occur for example as isolated stocks in the hard limestones at the base of the Sphinx. Further, AIGNER observed that landwards lagoon sediments follow, poorer in nummulites, but richer in fossils, with continental influx, forming in parts low, sandy, thick banks of a limestone/marl sequence, which is also observable on the body of the Sphinx. The ARCE group classifies the entire sequence below the head of the Sphinx as the Mokattam formation, which was then subdivided by GAURI[67] into three members: The base is formed out of the Rosetau Member (named after the ancient Egyptian designation of the 18th dynasty enclosure wall of the Sphinx). These rocks consist mainly of bio-detritus that fill up the interstitial space between the above-mentioned isolated coral stocks, giving the sediment surface an irregular relief. Above it follows the Seteped Member (named after the New Kingdom name of the Sphinx sanctuary). This almost 10 m thick series is formed of about six 1–2 m thick limy and more marly layers that show a gradual decrease in salt content from 3.5 % to only 1.5 % at the top. The salt mainly consists of halite (NaCl), whereas gypsum and various potassium salts are rare[68].

According to the ARCE group, the uppermost part of the Mokattam formation in the Gizeh area is the Akhet Member (named after the ancient Egyptian word for horizon; Akhet-Khufu was also the name for the entire pyramid plateau). This Akhet Member forms the upper dorsum and especially the neck and head of the Sphinx. It is roughly 9 m thick and consists in its lower part of a rather soft limestone, rich in clastic material. The geological properties of this member, particularly as it unfortunately occurs in the neck area of the Sphinx, cause the well-known problem of instability.

In contrast, the Ain Shams group is of the opinion that the stratigraphy of the Sphinx is analogous to the Observatory formation of the Eastern Desert mountain range, which extends from the south-eastern area of Cairo to Helwan.

The Ain Shams group has the advantage of a clear stratigraphic division of the Gizeh banks, dipping slightly about 6° towards the south-east. A similar stratigraphy is also given by AIGNER[69], however with some facies interfingering, which could, to some extent, explain the above-mentioned discrepancies.

Basic information on the subdivision of the Eocene in Egypt is given by STROUGO[70]. YEHIA's geological map[71] shows a more detailed differentiation of the Mokattam formation at the Gizeh plateau, especially its western continuation. According to him, the actual basement of the pyramids is formed by sequences of the Upper Mokattam formation. Below them, dark grey, hard dolomite limestones of the Middle Mokattam formation overlay white to yellowish limestones with nummulites belonging to the Lower Mokattam formation.

The Upper Eocene of the Gizeh area belongs, according to the ARCE group, to the Maadi formation, which is rather heterogeneous and which AIGNER[72] describes genetically as a marine regression, starting in the Upper Mokattam and continuing until the Upper Eocene Maadi formation. The rocks of this unit consist of fossil-poor to fossil-free marly limestones, marl and sandstone horizons, and shell beds. The Maadi formation generally changes in facies and thickness over short distances, as is usual for a very shallow marine environment. Prominent banks in the Upper Maadi formation are the Ain Musa beds with their hard and fossil-rich limestone banks. Later, during the Pliocene marine transgression, the softer Maadi beds were washed away, causing a collapse of the Ain Musa beds. Today, these collapsed beds form the hill-large blocks to the south and west of the Islamic cemetery situated south of the Sphinx, and were given the local name Hitan el-Gurob.

Finally, YEHIA mentions sediments of the sandy-conglomeratic Oligocene Gebel Ahmar formation occurring west of the Mena House hotel and west of the Fayum road[73]. In contrast with the type locality, the sediments do not contain silicified quartzite, which makes the stone suitable for use in construction.

The tectonical history of the Gizeh plateau was most probably of great significance to the decision of where to build the Great Pyramids. This becomes obvious by the field investigations of YEHIA[74], who mapped and interpreted the main fault lineaments, yet without discussing the possible consequences for the choice of pyramid site.

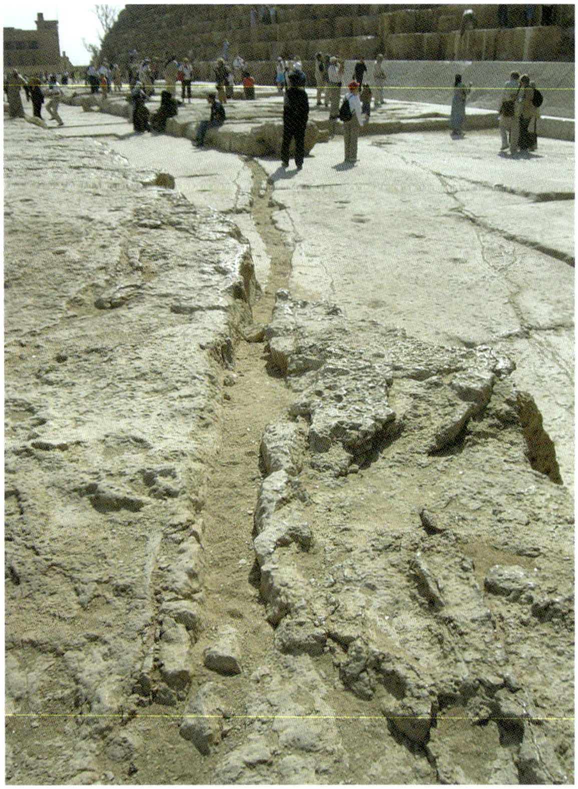

Fig. 72: F1-fault system running NNE-SSW through the Gizeh plateau, here towards the Khufu pyramid.

According to him, a (geologically) young F1-fault system runs in a NNE-SSW direction for approx. 250 m behind the western side of the Mena House towards the Khafre pyramid without reaching it, but replacing a Quaternary Nile terrace. A number of other smaller faults run in that direction, directly into the rock base of the Khufu pyramid (Fig. 72). This feature had certainly been recognized by the pyramid engineers, as steps were taken to incorporate it into the planning: It seems that the base length of the Khafre pyramid was reduced from the originally planned 60 cubits (274 m). Further, the shift of the entrance of more than 14 m from the central axis could be regarded as a consequence of this geological situation. If the entrance passage would be constructed along the real symmetric middle axis, the burial chamber would come dangerously close to this fault system, thus resulting most probably in problems concerning static. A drastic increase of the pyramid's weight, on a base of 60 cubits in length, was of course far too risky.

The second F2-fault system was of great importance to the architects in charge of the Menkaure pyramid, as it runs diagonally, almost directly through the pyramid's rock base. This might be the reason that the project's volume was kept to a minimum and was most probably originally planned to be even smaller. The position of the Menkaure pyramid at its present site, however, was determined by the overall planning of the Gizeh plateau[75]. Further, the construction of the pyramid towards the south-west away from the fault system would have made the addition of satellite pyramids problematic, as one would have left the southern limit of the hard nummulitic limestone units of the Upper Mokattam formation within the Gizeh plateau.

The other fault systems mentioned by YEHIA are of no direct influence to the Gizeh monuments and must not be discussed here in detail. Furthermore, AIGNER observed the special tectonic feature that the northern edge of the Gizeh plateau was controlled tectonically and later eroded by Pliocene marine erosion[76]. Incidentally, this edge runs parallel to YEHIA's fault systems F2–F6[77].

The Gizeh Quarries

1. The main quarry area, supplying the core masonry of the Khufu pyramid, was situated some 500 m south of the pyramid's southern edge (Fig. 73, 73a-b). Modern satellite images show evidence of a drag ramp running from the western part of this quarry area towards the south-western corner of the Khufu pyramid (Fig. 74). In fact, recently, while laying an electrical cable, the remains of two parallel narrow ramps were discovered that lead to the south-western corner[78]. This ramp was later overbuilt by the Khafre causeway, which was then used as a drag ramp during the construction of his pyramid. Studying Fig. 74, a second drag ramp runs from the eastern side of the Khufu pyramid, bending slightly to the west into the quarry area. This second ramp was also overbuilt by the Khafre causeway, which is thus younger than the ramp. In accordance with LEHNER[79], this site is referred to as the Khufu quarry in the

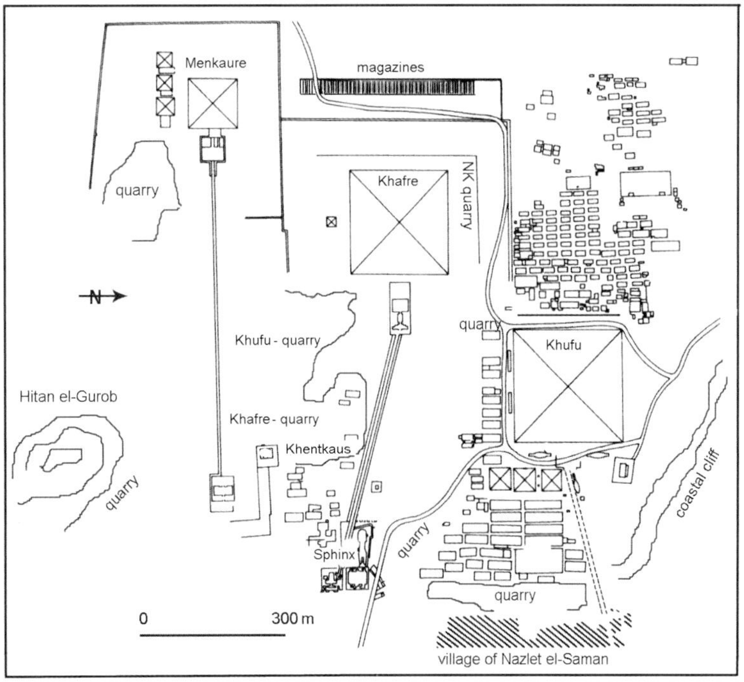

Fig. 73: Sketch map of the Gizeh plateau with the sampled quarry sites.

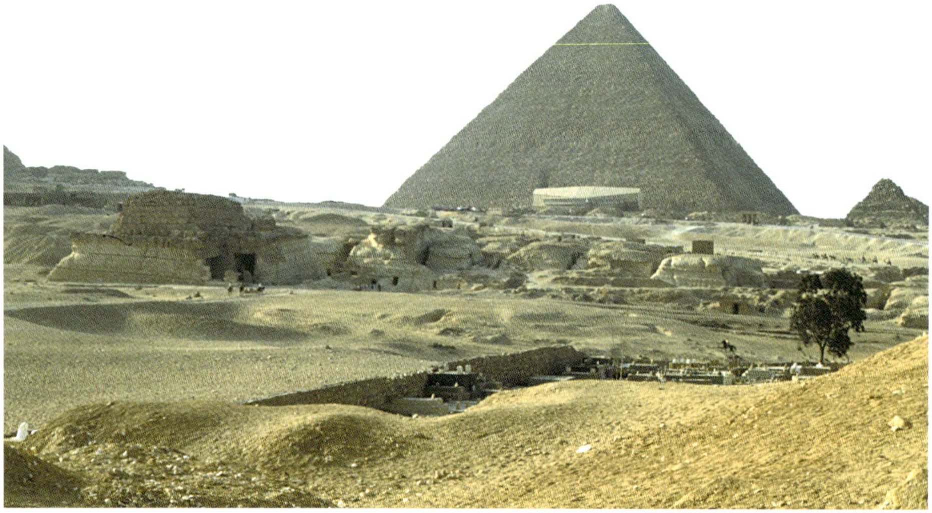

Fig. 73a: Main quarry site of Khufu and of Khafre. The excavation is still visible at the rims, but the filling of the large open pit with blown sand made an early identification as a quarry difficult.

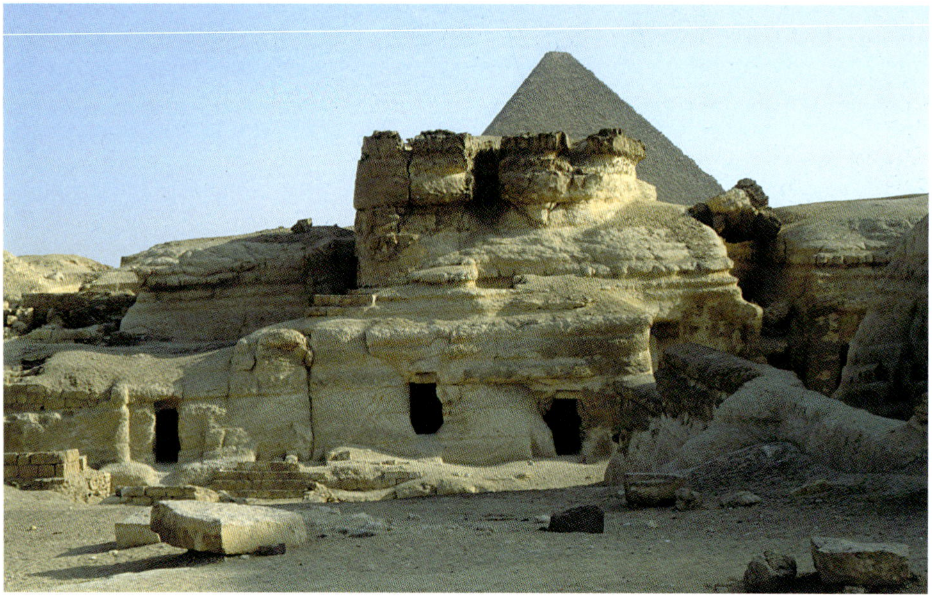

Fig. 73b: Western part of the main quarry site preferently worked by Khufu. Dark openings are later excavated rock tombs.

present monograph and is regarded as the western part of the central quarry field. Later, rock tombs of Khafre's family and of the 5th dynasty were cut into the quarry walls[80]. The more eastern parts of this central quarry field were generally exploited by Khafre to gain core material for his pyramid. This quarry was then extended up to the area where the Great Sphinx was formed[81].

REISNER identified this main quarry on the Gizeh plateau as belonging to Khufu and Khafre[82]. It also includes the area of Khentkaus' mastaba, which is built over and around a block of massive stone left by the quarrymen.

Macroscopically, these limestones are of a grey-beige to yellow-brown colour, mostly compact but also porous in places, and they feel chalky due to marly components. Many of small-sized fossil remains are detectable but hard to identify. Occasionally, small nummulites up to 5 mm in length could be recognized at polished surfaces. During storage over a longer period, various salts effloresce at the surface, which can be washed off easily with the finger.

With a hand lens, the fossils appear mostly as small nummulites, shells and other fossil remains, all irregularly imbedded and mostly secondarily calcified within the limestone matrix.

Under microscope, the rich variations of fossils and fossil fragments dominate the typical bioclastic rock structure. In spite of the macroscopic colour variation, under microscope all the rock specimens are similar. Apart from the main fossils as represented in Figs. 75–78, many other fossil remains such as oyster shells, echinoderms, sponges, and

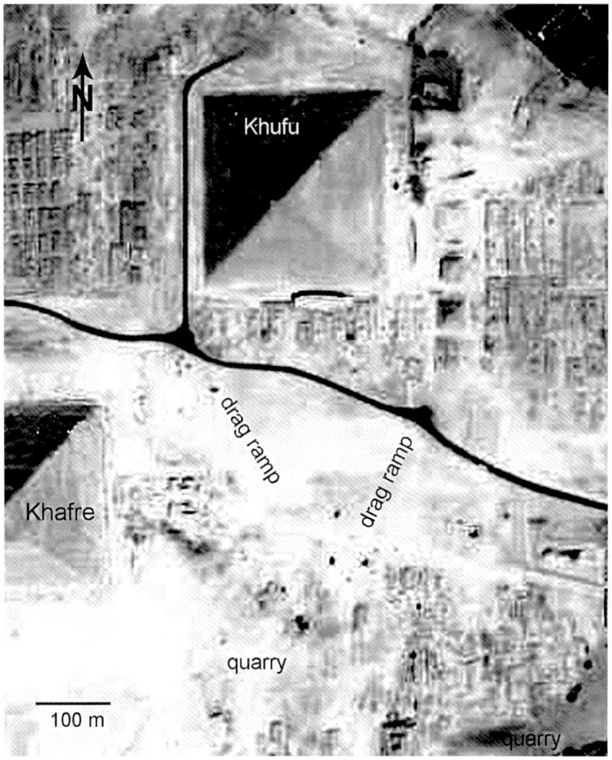

Fig. 74: Drag ramps running from Khufu main quarry site towards the south-west and south-east pyramid corners (mapped after Google satellite image 2006).

nano-fossils occur in the clay-rich limestone matrix and characteristically darken the optical impression. Rare, mostly sharp-edged quartz sand grains indicate a not too distant coastal line.

2. The steep escarpment east and north-east of the Khufu pyramid, close to the village of Nazlet es-Saman, was partly formed artificially by quarrying the rock there. The rectangular block characteristic of quarry structures can be recognized clearly in the upper parts, and appears on aerial photographs in an abnormally straight line along the

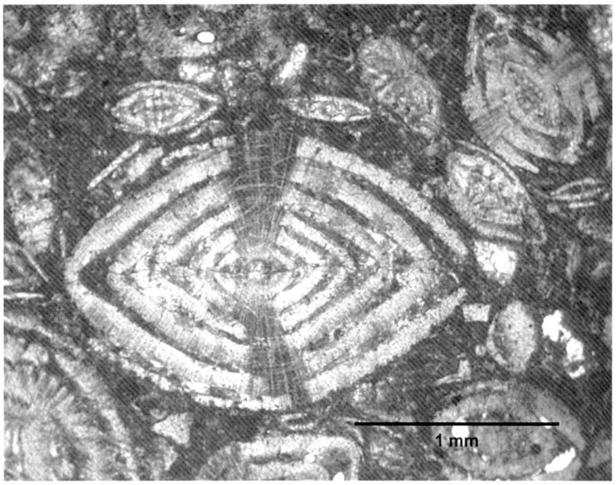

Fig. 75: Micro-photo (QS 1381a). Nummulites in microsparitic clay rich (dark) matrix. Main quarry of Khufu, south of his pyramid in Gizeh (1 pol. filter).

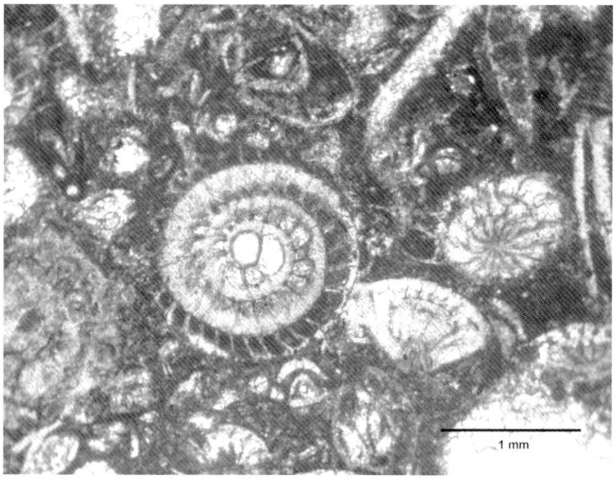

Fig. 76: Micro-photo (QS 1381b). Gastropode in fossil-rich microsparitic matrix. Main quarry of Khufu south of his pyramid in Gizeh (1 pol. filter).

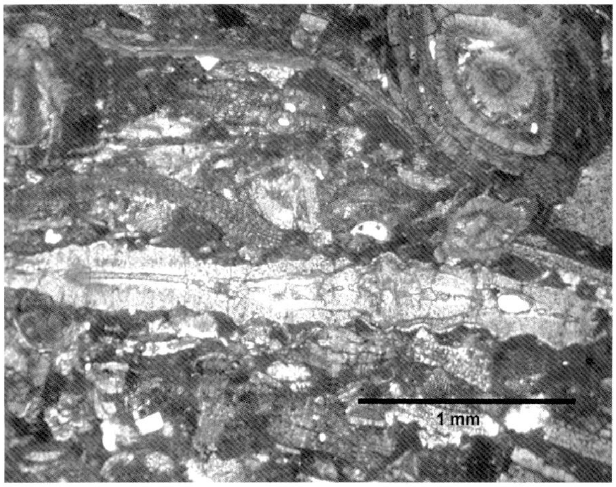

Fig. 77: Micro-photo (QS 1379). Dyscocyclina with nummulites and algaes in microsparitic-clayish matrix. Main quarry of Khufu south of his pyramid in Gizeh (1 pol. filter).

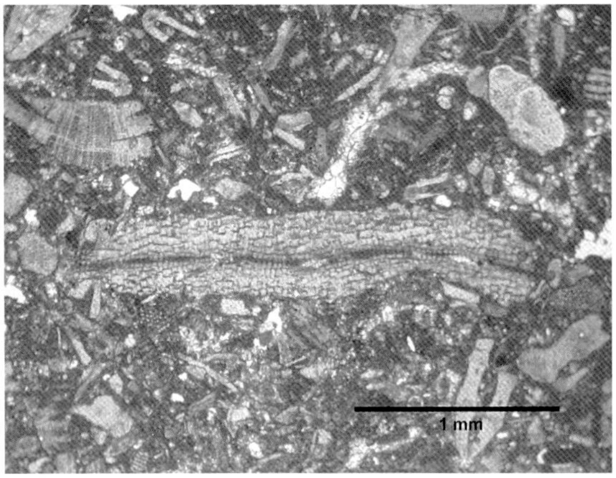

Fig. 78: Micro-photo (QS 1368). Algae with fossil fragments in microsparitic matrix. Main quarry of Khufu south of his pyramid in Gizeh (1 pol. filter).

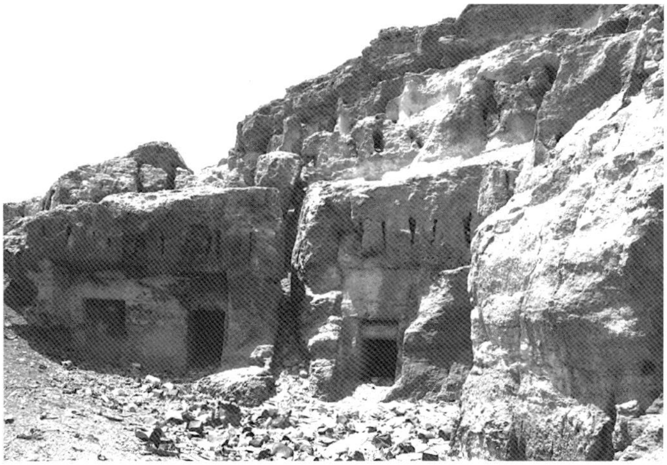

Fig. 78a: Detail of Khufu's eastern escarpment quarry with rock tombs from the 5th and 6th dynasties. The causeway was used first as a drag ramp for blocks gained from here.

escarpment's border. Presumably, part of the eastern rock slope was completely quarried away and the drag ramp was later used as causeway. A number of small rock tombs from the 5th and 6th dynasties that were cut into the remaining quarry walls can still be found there (Fig. 78a).

This geological unit continues until the base of the Khufu pyramid, where it crops out at various points of the entire platform. It was incorporated into the pyramid's body to quite an extent as can be observed at the southern side and in its internal chambers and corridors.

3. Further, the rocks obtained during the leveling of the rocky plateau were also used for the core masonry. Abundant traces of this quarry-type work are still clearly recognizable around the pyramids (Fig. 79–79a), and blocks of a considerable height can still be seen cut into the bedrock at the south-west corner of the Khufu pyramid (Fig. 79b).

According to Reisner[83], the bedrock of the eastern and western mastaba fields were also used as quarries. However, Hawass doubts that these mastaba fields were originally used for quarrying as the tombs had already been established during the reign of Khufu. Nevertheless, traces of quarries can be found around the mastaba fields and even between the (individual) mastabas[84].

Reisner also regarded the plateau to the west of the Khafre pyramid as a major quarry site. This is hard to confirm today.

Further, according to Reisner, "just north of the First Pyramid the edge of the rock platform can be followed roughly and appears to have been quarried back, but the scarp is now covered with a mass of masons' debris thrown out when the pyramid enceinte was cleaned up after the construction of the First Pyramid". This is partly in contrast to Aigner[85], who regards this part of the escarpment as natural Pliocene marine cliff erosion.

Finally, the rock-cut trench west and north of the Khafre pyramid yielded an enormous amount of stone material, which was incorporated directly into the core masonry.

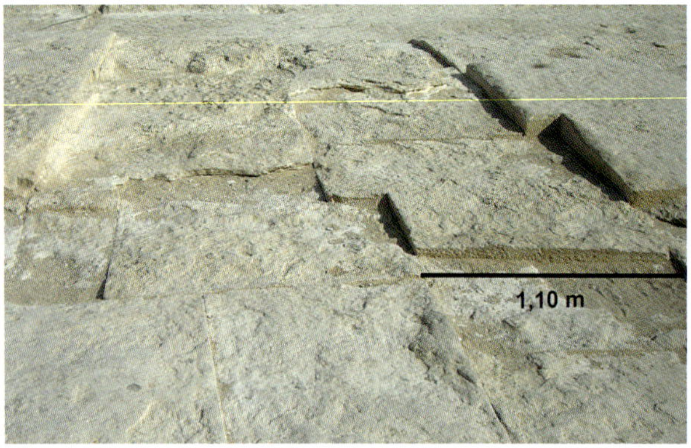

Fig. 79: Northern plateau of Khufu pyramid with traces of quarry-type leveling.

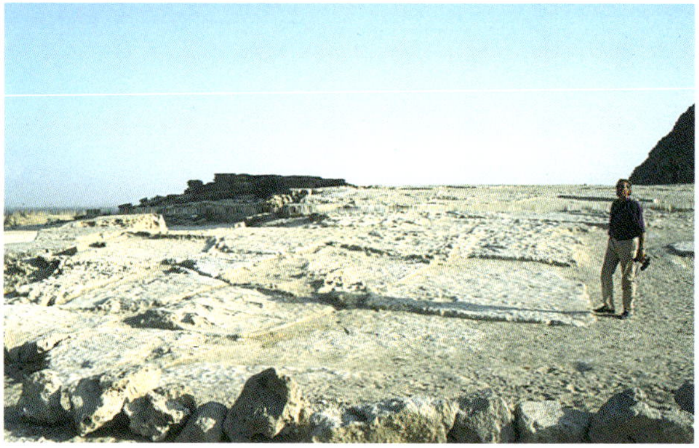

Fig. 79a: Plateau east of Khafre pyramid. The stone material gained by leveling the plateau was used for pyramid masonry.

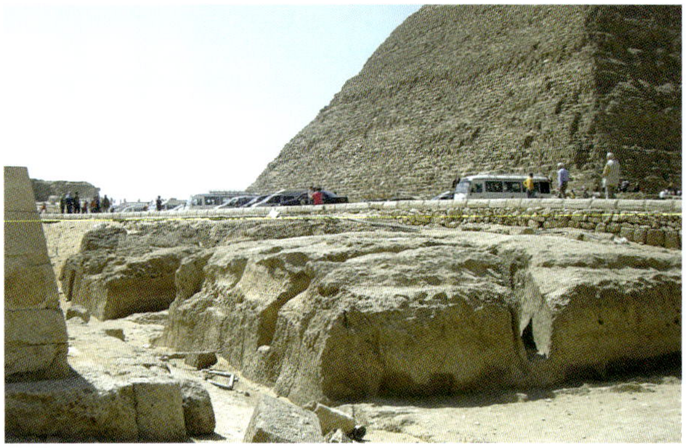

Fig. 79b: Remaining quarry structures at the south-west corner of Khufu pyramid caused by leveling the plateau. Note the block sizes and the wedge holes at the base to take off the separated blocks.

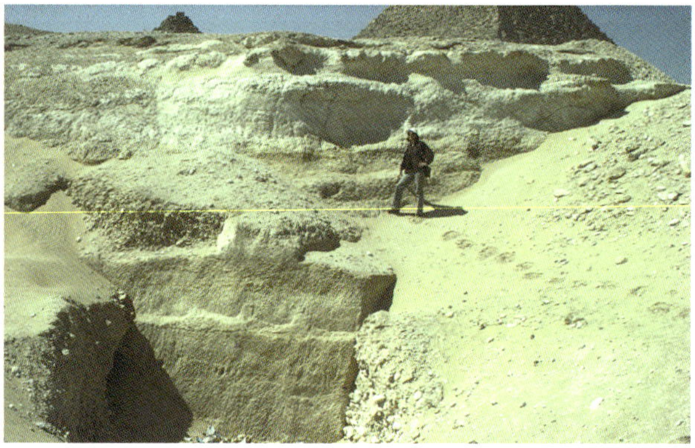

Fig. 79c: Menkaure quarry south-east of his pyramid. Well recognizable is the average height of the blocks gained from here.

4. In addition, clear evidence of quarry workings can be found at the southern edge of the Gizeh pyramid field along the northern flank of the modern Sphinx road, known as the *Route Touristique* (see Fig. 73).

5. To the south-east of the Menkaure pyramid lies an isolated quarry area, which was always regarded as the quarry for the Menkaure complex. The rock tombs there date mainly to the 5th dynasty, and, as will be shown later, this quarry is the only site from where the core material for the Menkaure pyramid was extracted (Fig. 79c).

The samples from these last sites differ to some extent from those of the great central quarry field, even when they belong geologically to the same unit. They appear coarser-grained and consist of highly calcified bioclastic to fossil-rich

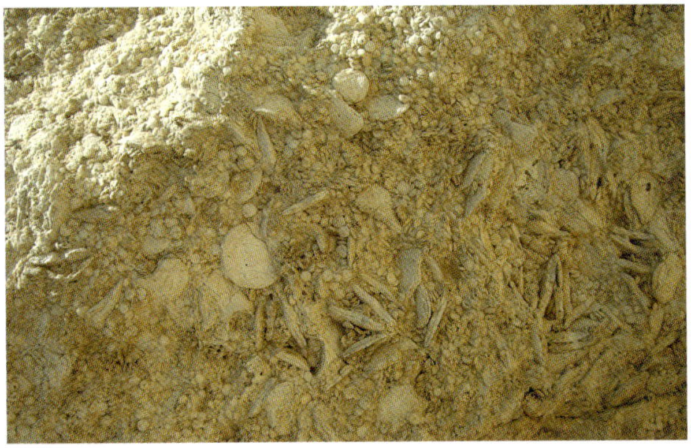

Fig. 80: Nummulite-rich limestone from the southern core of Khufu pyramid.

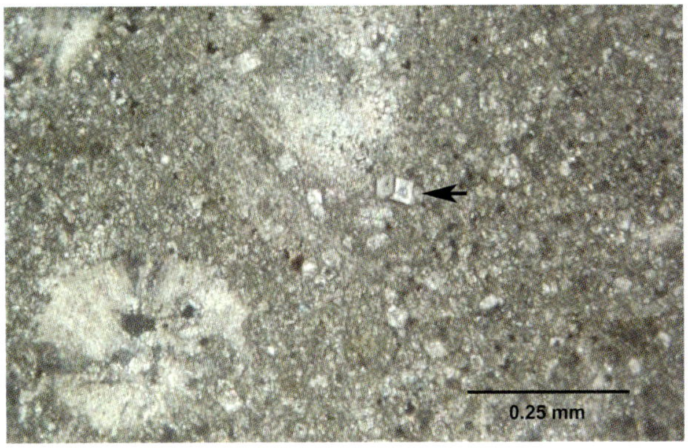

Fig. 81: Micro-photo (QS 1548) of limestone from escarpment quarry east of Khufu pyramid. The well formed dolomite crystals (arrow) in the relative fine-grained matrix are striking. Further, the small nummulites appear almost as ghost structures due to diagenetic dissolution (1 pol. filter).

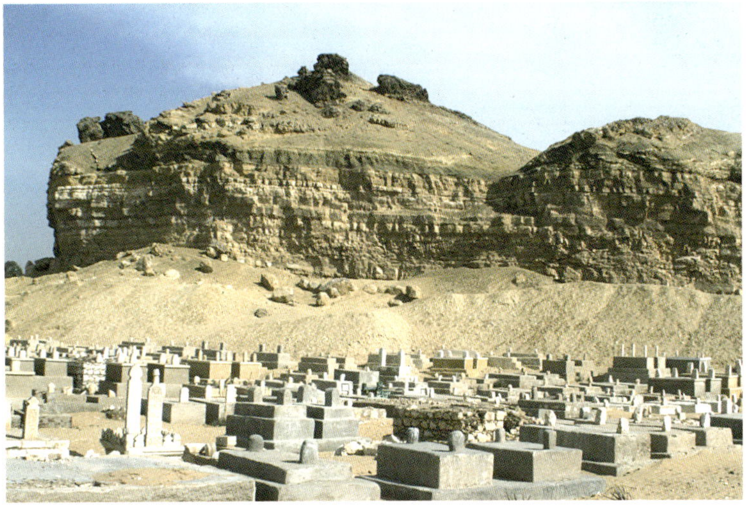

Fig. 81a: Hitan el-Gurob, a prominent hill south-east of the pyramid plateau, which was also a source for core material of Khufu and Khafre pyramids.

limestones of grey to greyish-beige colour. Some specimens even contain large *Nummulites gisehensis*, pored together, and various large shell fragments. After a period of storage, these limestones tend to effloresce mainly with halite.

Macroscopically, large nummulites are embedded in a groundmass comprised of smaller examples (Fig. 80). Herodotus reported this specific stone type as petrified lentil meals of the workers.

With a hand lens, the rich bioclastic debris and small nummulitic fossils in addition to the large *Nummulites gisehensis* are clearly recognizable. Approx. 0.5 mm-sized calcite crystal grains reflect the light with their cleavage planes.

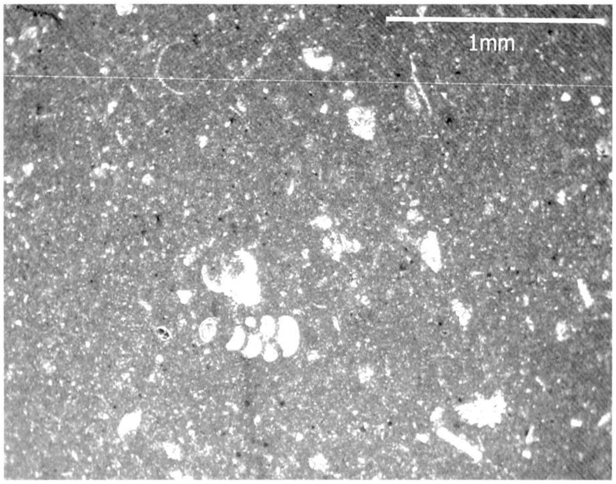

Fig. 82: Micro-photo (QS 1587) of Hitan el-Gurob limestone, much resembling the Saqqara-Meidum casing stones of the 3rd dynasty (compare Figs. 10, 34–35; 1 pol. filter).

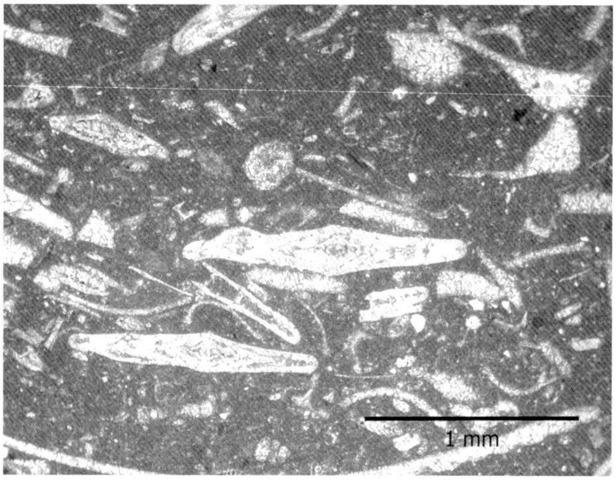

Fig. 83: Micro-photo (QS 1583) of Hitan el-Gurob. Fine-grained limestone with discocyclinae, remains of Nummulites and thin shells, all more or less calcified (1 pol. filter).

Under microscope, the samples taken from lower parts of the escarpment are equal to those from the great quarry. Samples taken from the upper parts differ from these significantly: The bioclasts and predominant nummulitic fossils are cemented by a coarser grained matrix with well-formed calcite and dolomite crystals in parts (Fig. 81).

6. A further quarry area that was also used by Khufu and Khafre could be identified at the outcropping rock walls of Hitan el-Gurob, some 800 m south-east of the pyramids and south of an Islamic cemetery (Fig. 81a).

Macroscopically, they consist of compact and hard bioclastic limestones to sand-rich limestones with a dense structure. A distinct colour variation from grey, yellow to almost reddish is characteristic. Visible fossil remains consist mainly of shell fragments but rarely nummulites.

With a hand lens, the shell fragments appear glassy due to intensive calcification. In contrast, small nummulites are much easier to recognize.

Under microscope, two main structure types are distinguishable: One with a very dense, microsparitic matrix with only few fossil remains of mainly small nummulites and globigerinae. This type is to some extent similar to the fine Meidum limestone and the casing limestones used in Saqqara during the 3rd dynasty (Fig. 82). The second type also consists of a very dense and fine-grained microsparitic groundmass but is dotted with fossil fragments, small nummulites, and discocyclina. Whilst these fossil remains mostly retain their original structure, the many broken shell fragments are always intensively calcified. Many of these samples contain within their matrix sand grains between 0.05–0.5 mm, mainly of quartz but also of plagioclase, thus indicating a certain degree of continental influx (Fig. 83).

Remarks on the discussion about pyramids built of poured concrete blocks

At this point a few remarks are necessary about a number of recent publications concerning the building material of the pyramids. DAVIDOVITS/MORRIS[86] published a passionately discussed book "The Pyramids: An Enigma Solved", where they concluded that the building stones of the pyramids were formed by using artificial concrete. According to this publication, the blocks consist of a cement mixture of a geopolymere and natural limestone aggregate, poured into moulds. From a petrographical point of view, they came to their conclusions based on one rock specimen, the so-called "Lauer sample", and a few rock specimens from the Gizeh pyramids. Despite the arguments proposed to substantiate their hypothesis, they could not convince Egyptologists and geoscientists who investigate Egyptian stones. Especially in the USA, DAVIDOVITS/MORRIS' thesis gave rise to a vehement scientific discussion. A critical response started with articles from internationally renowned geologists specializing in the sedimentology of limestones, who seriously doubted the geopolymeric hypothesis[87]. They give convincing sedimentological and structural arguments for a natural character and source of the pyramid building stones. HARRELL/PENROD came to the same conclusion[88], comparing the "Lauer sample" with samples from Gebel Mokattam, Tura, and Maasara. MORRIS[89] responded excitedly, discriminating HARRELL's results and declaring CAMPBELL/FOLK ignorant of the cast-in-place theory and the chemistry of geopolymerization.

After an intensive but fruitless discussion during the Cairo International Congress of Egyptology in 1988, the present author (D.D. KLEMM) invited JOSEPH DAVIDOVITS to the laboratory of the Department of Geosciences of the University of Munich. Here, a suitable collection of petrographic thin sections was mutually scrutinized under a polarizing microscope and the identity of stone samples from the pyramids and from the corresponding quarries was discussed. This demonstration seems to have ended the debate.

Nevertheless, BARSOUM/GANGULY/HUG[90] recently published a paper on the "Microstructural evidence of reconstituted limestone blocks in the Great Pyramids of Egypt", wherein they reopened the geopolymere-concrete discussion that has been debated so heatedly in the past few decades. BARSOUM/GANGULY/HUG introduced the term "microconstituents" (μc's), a term which denotes structural components that are impossible to detect with petrographic microscopy and visible only by applying scanning electron and transluminescent electron microscopy. Further, they presented a series of energy dispersive spectrometric (EDS) chemical analyses. The results show that the quarry samples contain natural minerals, for example calcite, dolomite, and silica. They also present very unconventional formulas of Si-Mg-Ca-C-oxy-hydroxy and in other cases Ca-K-Al-Si-C-oxy-hydroxy compositions from samples of casing stones from several Gizeh pyramids and declare them as artificial cast-in-place concrete-type stones. In contrast to DAVIDOVITS/MORRIS[91], they regard the core stones as natural in origin without giving additional arguments. Even though their analytical methods are explained correctly in an appendix, their results are hard to follow and are completely in contrast to observations of the present authors. Most probably, these unusual compositions have a simple explanation: The used acceleration voltage of 12 kV creates a volume of about 1–2 μm^3 at a well-polished surface orthogonal to the electron beam. However, within a broken granular volume with grain sizes below 1 μm, the electron beam also excites the submicroscopic fine-organic coatings of bituminous matter, always present in the various Egyptian limestones that are rich in organic fossil remains, like those from Gebel Mokattam, Tura, and Maasara. The structures analysed by BARSOUM/GANGULY/HUG are of this submicron size. Thus, the danger of a measurement contamination is very likely and would explain the exceptional results. Yet, these results are insufficient to prove the presence of geopolymere, as postulated by the authors.

In the present investigation, around 1500 samples from pyramids and quarry sites were studied using various petrographical and geochemical methods and, with the exclusion of only a few cases, the samples from the pyramid stones and respective quarry sites matched very well for both core and casing. A number of samples were analysed with scanning electron microscopy and EDS and compositions comparable to BARSOUM/GANGULY/HUG's findings could only be found in the cases of well-known minerals such as calcite, dolomite, quartz, feldspar, clay minerals, bituminous matter etc. Further, the volume of the detected quarry sites corresponds well with the volume of the pyramids constructed with this material. More importantly, natural rocks crushed down to aggregate size and then mixed with a binder to form a kind of concrete loose their original internal orientation. Nothing of that could be observed by the present authors, only well-preserved bedrock bedding and undisturbed secondary diagenetic structures with, if present, well-sorted fossils or their remains, in many cases calcified but free of any artificial binder. Most probably, the discussion about whether pyramids were constructed using geopolymere will continue. However, it should be clearly emphasized that such theories are nonsense.

1. The Pyramid of Khufu (2604–2581 BC)

Geographical position (present entrance): 29°58'48.3" N, 31°08'3.58" E

Height: 146.6 m | Base: 230.38 m | Slope: 51° 50'

Khufu most probably abandoned the royal necropolis at Dahshur because there was no longer enough limestone nearby and because of the low stability of the subsoil, which consists of slate clay. Instead, he decided to build his pyramid on a massive rocky plateau in the western desert near modern Gizeh (Fig. 84), where the subsoil was much

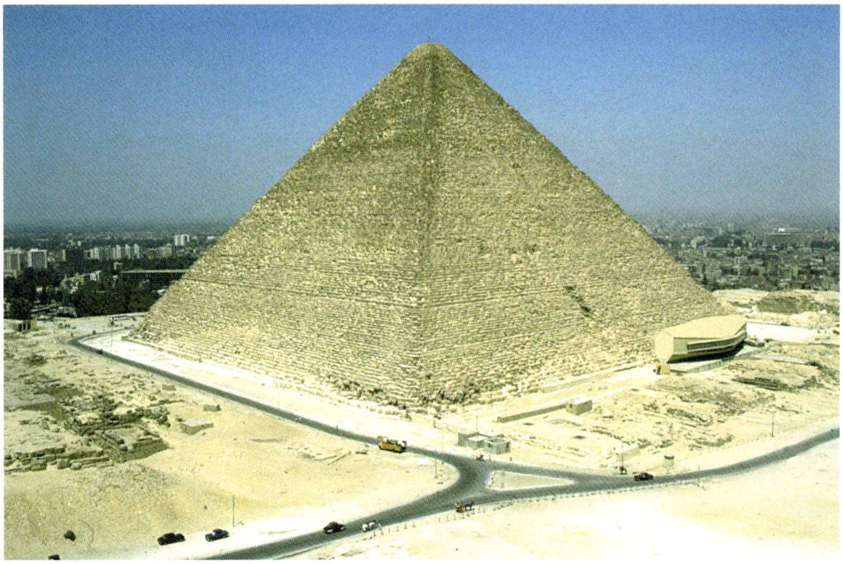

Fig. 84: Khufu pyramid from southwest, seen from top of Khafre pyramid.

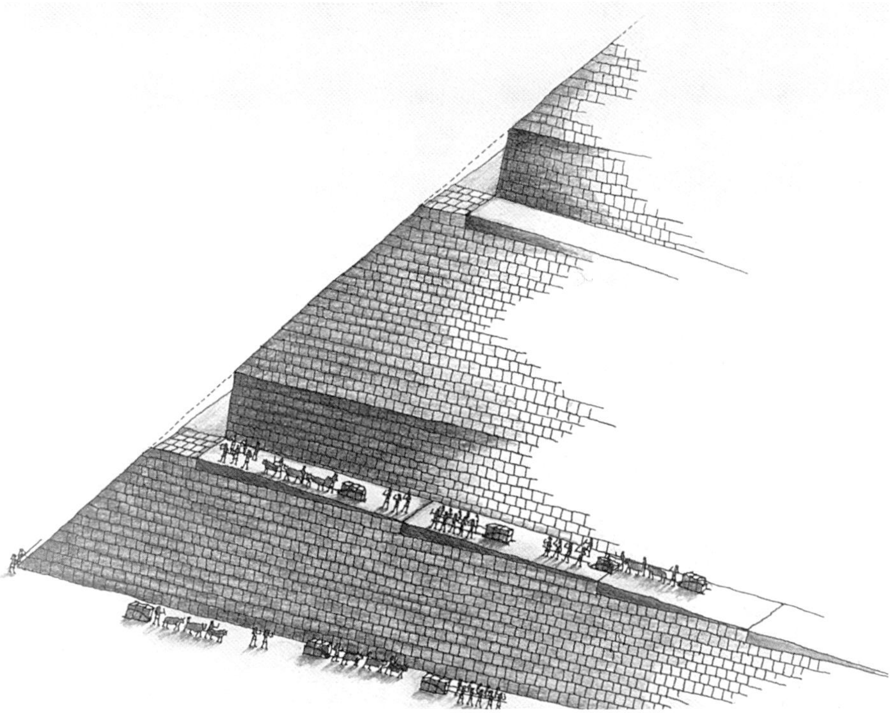

Fig. 85: Integrated spiral ramp for the construction of large pyramids in limited areas like the Gizeh plateau, thus avoiding enormous amounts of material commonly claimed for a separated outside construction ramp. Drawing by S. UHL.

more stable, and an abundant supply of high-quality limestone was available. In terms of size, technical accomplishment, and organization required for its construction, this new pyramid represents a phenomenal building.

The limestone used in the construction of the pyramid came from different quarries to the east and south of the building and, as demonstrated above, from the construction site itself. The limestone blocks were without doubt transported via drag ramps to the construction site. The pyramid was then probably erected by means of a system of ramps (a selection of potential models is presented by ARNOLD[92]).

A new model of spiral construction ramp at large pyramids such as the Khufu pyramid has been proposed by the present authors[93]. This spiral-type ramp is integrated into the pyramid's outer body, simultaneously leaving unfilled space of a suitable width while erecting the pyramid, and thus avoiding tremendous amounts of secondary materials like Nile mud, gravel, wood etc. to form a separate outer ramp.

When studying the Gizeh plateau and its surroundings by means of field observations, stereoscopic aerial photos, and satellite images, it is remarkable that there are no visible traces of waste dumps produced by such enormous constructions. This cannot be neglected, when postulating separately constructed ramps, as is often the case. An integrated ramp, however, which was later filled up with additional core and casing blocks while finishing the pyra-

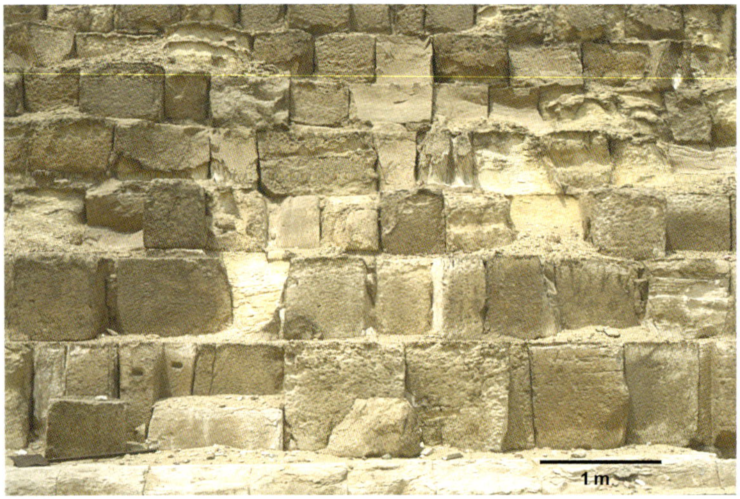

Fig. 86: Block sizes of Khufu pyramid. They differ slightly in width but they are equalized in height.

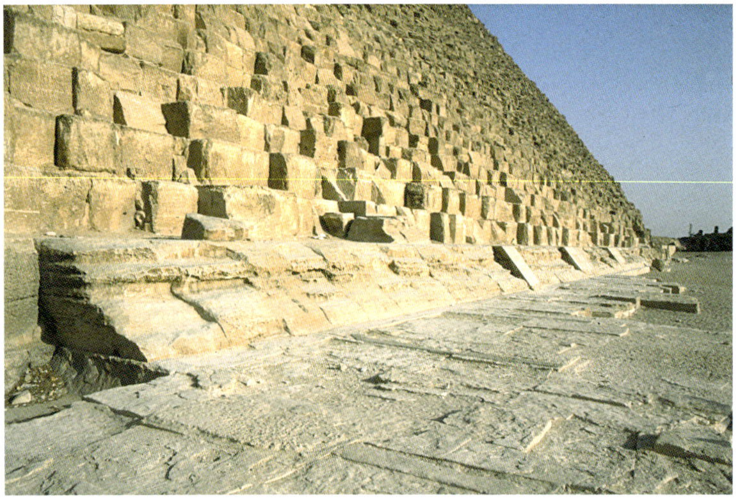

Fig. 87: Casing and backing blocks of Khufu pyramid, western side.

mid, has the advantage that large amounts of debris were not in the way at the construction site, nor did they have to be deposited elsewhere after the pyramid's completion. In addition, an integrated ramp allows a clear view on the angles for exact measurements during construction and enables the use of two opposite ramps for transport up to and down from the pyramid, thereby considerably accelerating the whole building process (Fig. 85).

The outer walls of the core are built with blocks laid in horizontal layers. The height of the blocks varies between 0.80 and 1.20 m in average (Fig. 86). Between the core and the casing, another layer of somewhat smaller stones was bound with mortar, which increased the cohesion of both the two materials and the two masonry structures. In archaeological terminology, this intermediate layer is known as "backing stones" (Fig. 87).

The casing was made of large blocks of fine, white limestone, but few blocks are still in place, mostly at the base. As in the case of the older Red Pyramid, slightly concave walls were intended to increase the stability of the pyramid's mantle.

A possible valley temple and an adjacent harbour have been located beneath the village of Nazlet es-Saman. Pieces of basalt pavement and limestone walls, most probably belonging to these structures, were unearthed in this area[94]. A causeway from Nazlet es-Saman leads up to the mortuary temple that was situated on the eastern side of the pyramid. A cult pyramid and three small queens' pyramids also stand at the south-eastern side.

All that remains of the mortuary temple east of the pyramid is a black basalt pavement, sockets for granite pillars of the surrounding colonnade, and a number of limestone bedrock cuttings for the outer walls.

Analytical data does not enable the differentiation of the various basalt outcrops in northern Egypt (i.e. Haddadin flow[95]), but on the basis of archaeological findings[96], it is most likely that the basalt slabs from the Khufu temple come from Widan el-Faras north of the Fayum lake. It is the first time in Egyptian architecture that basalt was used on such a large scale for paving. From this point on, this innovation was repeated by the following pharaohs[97]. Traces of sawing the basalt slabs using a drag saw can be seen on several pavement blocks[98].

The walls of the temple were of fine limestone and they must have been carved in relief, but only few decorated fragments have been found. Some blocks were later reused as building material in Amenemhet I's pyramid complex at Lisht[99].

The mortuary temple had a rectangular plan some 52.5 m wide. Hence, it was much larger than the small temples adjacent to earlier pyramids.

The original entrance with the descending corridor is about 17 m high starting on the level of the 13th layer at the north side of the pyramid.

In the burial chamber, a red granite sarcophagus is still *in situ*, oriented north-south. It is 2.24 m long and 0.96 m wide. The lid of the sarcophagus is missing. This large sarcophagus was installed during the construction of the granite chamber. Pressure-relieving chambers above the burial chamber are also made of large granite blocks.

Recently, Salah el-Naggar[100] presented a compilation of all granite components of the Gizeh pyramids and localized them all to the large granite and granodiorite occurrences south of Aswan.

Maragioglio/Rinaldi[101] measured the pyramid of Khufu and most of the other Old Kingdom pyramids, and presented their results in detailed plans.

Geochemical provenance determination of the core material

The correlation diagrams of Figs. 88–91, showing the geochemical distribution fields of the samples taken from the Khufu pyramid and surrounding quarry areas, attest to the fact that the main core material mainly derived from the quarry areas to the south and south-east of the pyramid. This is astonishing, as mainly rock tombs established during Khafre's reign are exposed in this area[102]. However, it should be remembered that Khufu granted his family members and high officials the privilege to build mastaba tombs to the west and east of his pyramid. Thus, the excavation of rock tombs in the walls of his main pyramid quarry was unnecessary. For the first time under Khafre, the necropolis organization with regards to the high nobility changed, obviously due to the better-exposed quarry walls that had been created in the meantime and also due to the proximity of his pyramid.

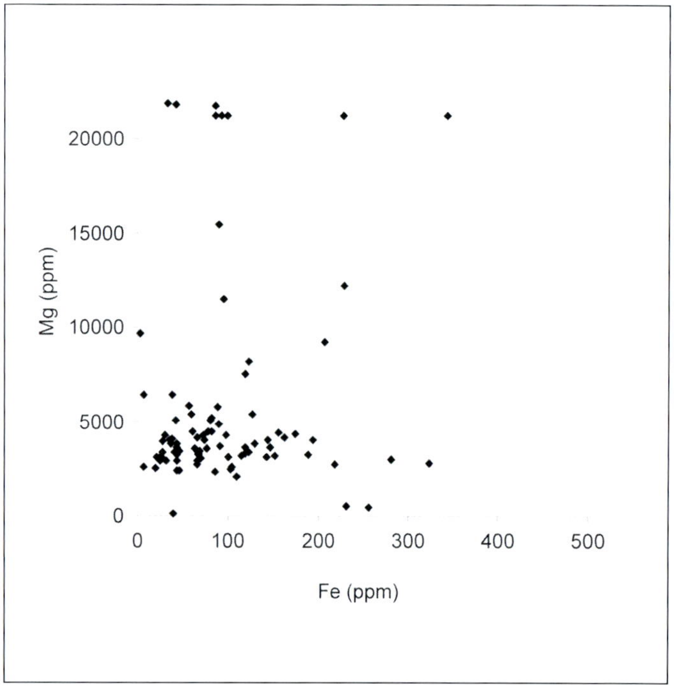

Fig. 88: Mg/Fe-diagram of core masonry from Khufu pyramid. A relatively wide distribution field indicates various sources of the block material.

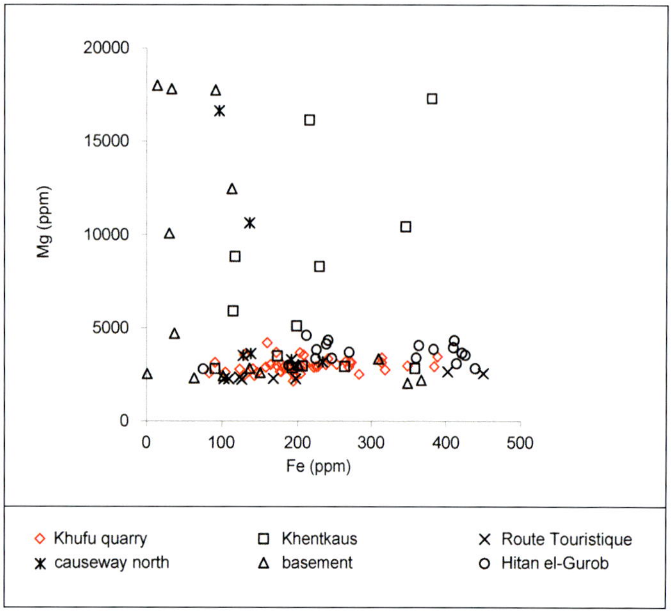

Fig. 89: Mg/Fe-diagram of the Gizeh quarry sites assumed to be the sources of Khufu's core masonry. The maximum of the block material derived from the quarries south of the pyramid and from the bedrock of the pyramid plateau (see sketch map Fig. 73).

Fig. 90: Mg/Sr-diagram of core masonry from Khufu pyramid. Again, the core data cluster within a limited field, corresponding with the southern quarries, but other sites like Hitan el-Gurob and the pyramid basement are obviously also source candidates.

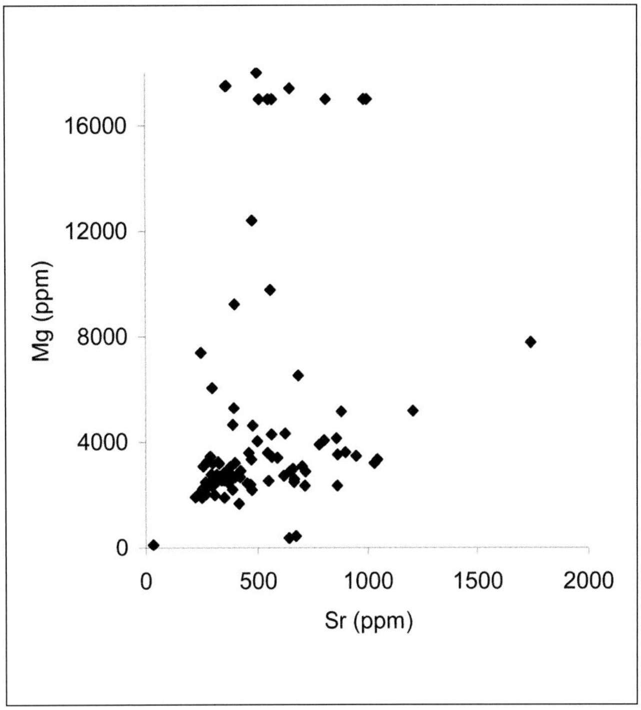

Further, the Mg/Fe-values of samples from the basal core masonry (Fig. 88) cluster in a similar field as the basement rocks (Fig. 89). This may indicate that quite a large amount of the material gained during basement leveling (see Figs. 79–79a) was directly used for the pyramid's construction. As a relatively large amount of samples was taken at the lower levels of the pyramid to achieve a statistical average, this correspondence is not surprising.

The geochemical diagrams for the core material (Figs. 88, 90) and quarry samples (Figs. 89, 91) also indicate that the area of Hitan el-Gurob served to some extent as a source for the material of the core masonry. This source had not been considered hitherto, but both the geochemical and petrographical investigations unambiguously come to the same result.

Quite an amount seems to derive from a quarry area that forms the escarpment north of the present *Route Touristique* and from the quarries at the escarpment south of the Khufu causeway.

Geochemical provenance determination of the casing and backing material

Concerning the casing masonry of the Khufu pyramid, a fundamental problem arises in that only a small number of actual casing blocks exists at the pyramid's base along with a few scattered blocks on the pyramid's plateau. Nevertheless, it is possible that these blocks did not originally belong to the structure but to the material of other pyramid complexes. Consequently, they were not sampled systematically.

The casing material consists of a whitish-grey to whitish-yellow, very fine-grained and dense appearing limestone and can be easily distinguished even by a less petrographically trained observer from the heterogeneous core masonry with its much coarser structure.

The material of the so-called backing masonry, which normally forms the two blocks behind the casing, is similar in both appearance and macroscropically to the remains of the casing blocks. However, the number of layers of backing blocks is irregular and can constitute up to four courses between the casing and the core masonry.

The uppermost 10 stone layers of the remaining structure seem to consist exclusively of backing masonry, and the core masonry there is not exposed. Thus, from 75 samples of casing and backing masonry, only about 10 are from the actual casing itself.

The geochemical correlation diagrams Figs. 92–93 display the plotted data from samples of the Khufu pyramid's casing and backing material with the quarry samples of Tura, Maasara, and Mokattam. At first glance, the Khufu casing and backing stones can be geochemically attributed to the Tura-Maasara limestones. An attribution to the upper layers of the Mokattam quarry district is also possible. The petrographical study shows that Maasara is a less likely source. The diagrams show that all casing and backing samples from the Khufu pyramid correspond well with these three provenance areas. Particularly when analysing the Sr-data, a number of the samples derive from Mokattam, but the greater part corresponds better with the Tura-field.

Finally, a second analysis of the sample protocols showed, in accordance with the sample collection, that there is no significant difference between the provenance of casing and backing material for the Khufu pyramid.

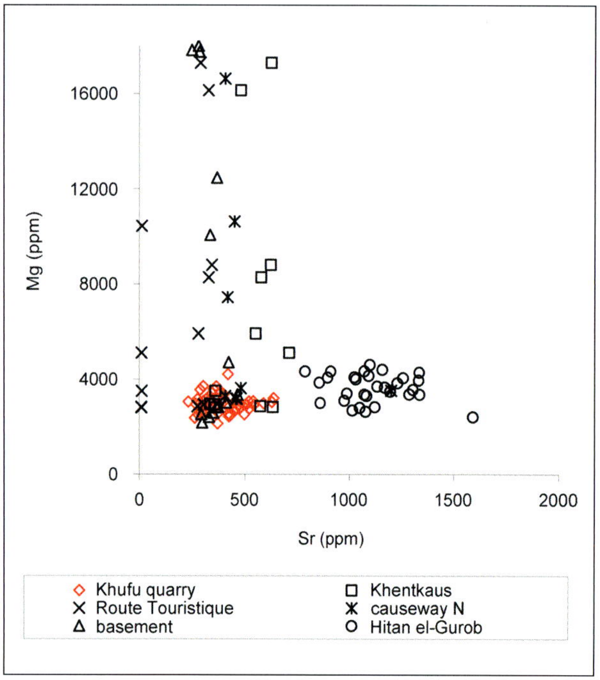

Fig. 91: Mg/Sr-diagram of the Gizeh quarry sites, assumed to be the sources of Khufu core masonry. Predominantly the great quarry sites south of Khufu pyramid and the pyramid basement turn out as the main sources for the core masonry.

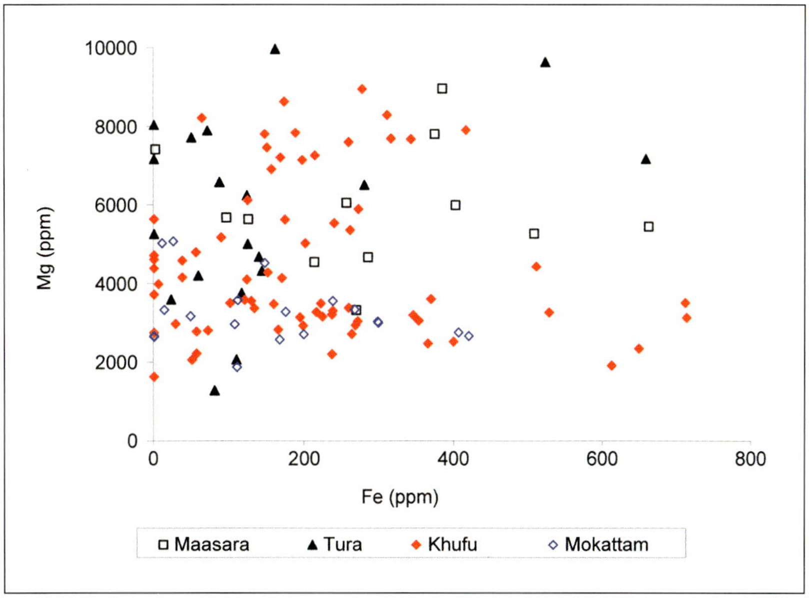

Fig. 92: Mg/Fe-diagram of casing and backing masonry from Khufu pyramid and the assumed quarry sites.

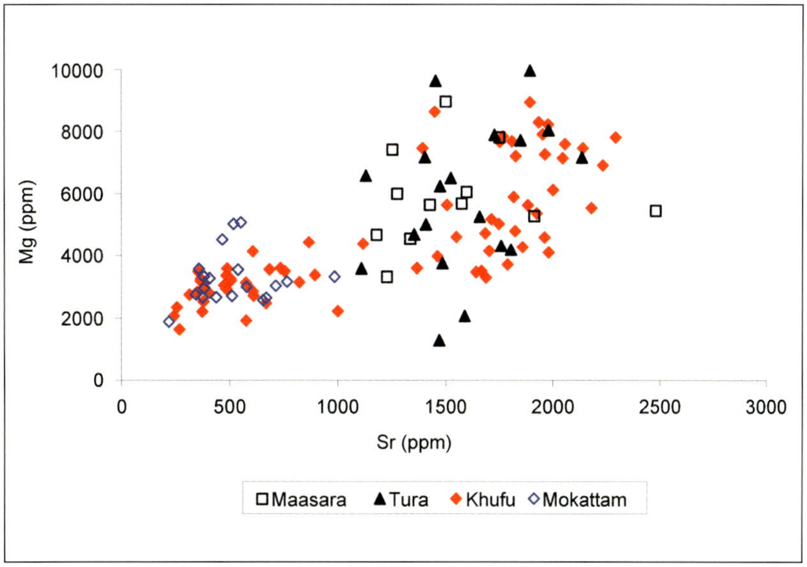

Fig. 93: Mg/Sr-diagram of casing and backing masonry from the Khufu pyramid and the assumed quarry sites.

2. *The Pyramid of Khafre (2572–2546 BC)*

Geographical position (present entrance): 29°58'36.67" N, 31°07'50.59" E

Original height: 147.5 m | Base: 215.25 m | Slope: 53° | Length of causeway: 494.6 m

The pyramid (Fig. 94) is 10 m higher in relation to the Khufu monument and thus, depending on the viewing position, seems to be bigger. The geological environment is almost identical to that of Khufu and does not need further discussion, all the more, as the slight geological variations in stratigraphy are of no significant importance. This is especially the case for the core material, which derives from the same geological level as the greater part of the building stones of the Khufu pyramid. Nevertheless, there is some material variation at the base of the Khafre pyramid, which is cut into an outcropping slope of the bedrock that forms the highest point of the nearby hilly landscape.

Like the pyramid of Khufu, the Khafre pyramid also takes advantage of outcropping rock, which increased the stability of its core and decreased the amount of building material. The bedrock surface to the north-west had to be cut down some 10 m, while the south-east corner had to be built up using huge blocks of masonry.

During the New Kingdom, the north-west corner of the original precinct was enlarged by a quarry, which was cut into the original rock most probably for restoration purposes. The traces of ancient quarrying can still clearly be seen there (Fig. 95). The rock walls in that corner are significantly less deteriorated than the original wall of the western side. A rock inscription of Maja on its northern face dates this quarry to the reign of Ramses II.

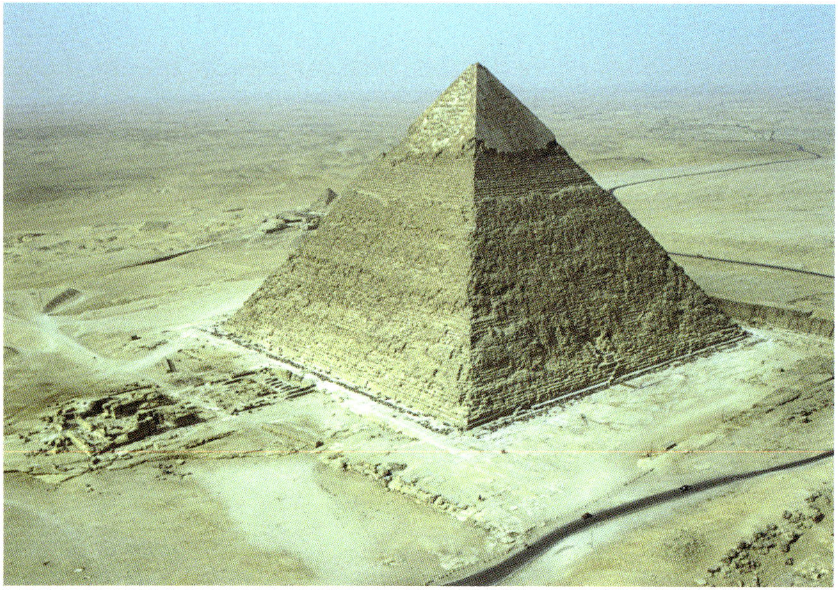

Fig. 94: Khafre pyramid with its cap of remaining casing, seen from top of Khufu pyramid.

The lower levels of the pyramid's casing were lined with pink granite, while the higher layers, which become smaller towards the top, are of fine limestone. At the top of the pyramid a small portion of the original casing remains in place, which allows us to see how the final blocks were laid and bound to the pyramid core (Fig. 96).

The pyramid is surrounded by a stone perimeter wall, encompassing an open courtyard around the pyramid. This courtyard is paved with limestone slabs of irregular form.

A small cult pyramid stands on the structure's central axis at the southern side of the pyramid.

The older of the two entrances into the subterranean rooms of the pyramid is located about 30 m north of the pyramid and carved completely out of the subsoil rock. The second entrance is located in the north side of the pyramid about 12 m above ground level. It meets a corridor lined with red granite that first descends into the pyramid before running horizontally at the structure's base. The burial chamber, east-west oriented, was excavated completely out of the bedrock. Near its west wall stands a black granodiorite sarcophagus that was originally covered by a sliding lid. The lid was found nearby in two pieces.

Finally, an enclosure wall may have extended around the entire Khafre pyramid complex, even including the Great Sphinx.

The valley temple of the Khafre complex, which is one of the best-preserved Old Kingdom temples, was fronted on the east by a large terrace paved with limestone slabs. The temple was laid out in an almost square ground plan and is situated next to the Great Sphinx and its associated temple. Its core wall was built of huge blocks of limestone, extracted from the nearby quarries around the Sphinx[103]. This inner core was then covered by granite slabs of various types, giving the temple the name "granite temple"(Fig. 97). All of the granite varieties used in the valley temple derive from the Aswan area. Between the two entrances of the valley temple was a vestibule with walls of red and greyish-red granite that were originally polished. Its floors were paved with white calcite-alabaster. A door then led into a hall, which was covered with polished red granite and paved also with white alabaster. It was adorned with sixteen

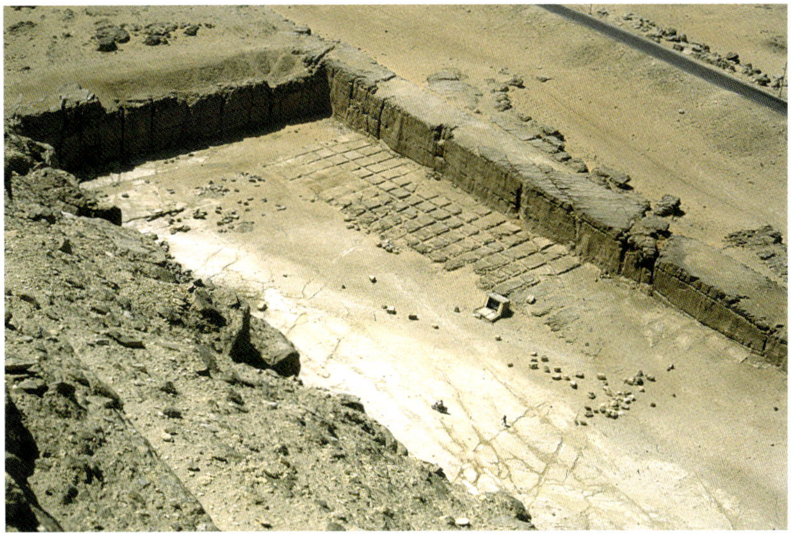

Fig. 95: New Kingdom enlargement of Khafre northern enclosure trench, most probably for gaining stone material for restoration purposes. Seen from top of the pyramid.

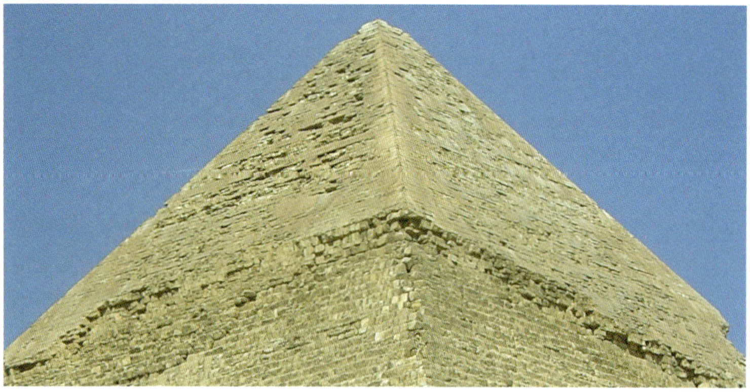

Fig. 96: Khafre pyramid. Remaining casing and backing blocks at the cap.

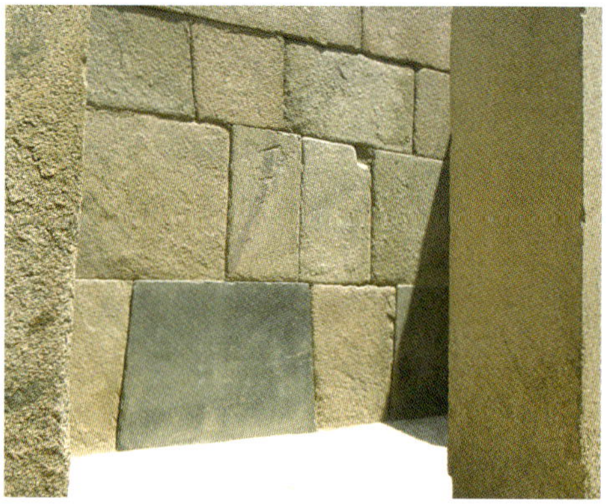

Fig. 97: Detail of the valley temple of Khafre, entirely faced with different granite varieties from Aswan.

red granite pillars, many of which are still in place today. The pillars supported architrave blocks of the same material. Again, this granite material also comes from Aswan. Statues of the king made from anorthosite, schist, and calcite-alabaster once stood here. The famous "Khafre with the falcon" is now exhibited in the Cairo Museum. Its stone material, whose petrological nomenclature is anorthosite gneiss, was commonly called "Khafre Gneiss" according to the gneiss-like appearance of this statue. This type of stone comes from the remote Gebel el-Asr area, some 30 km west of Toshka and around 250 km south of Aswan.

The mortuary temple, unlike in later pyramid complexes, did not directly border the pyramid but was separated from it by a courtyard. Its walls are built of local limestone cased with fine limestone, a technique newly introduced in this building. Inside, it was almost completely lined with granite (Fig. 98).

In contrast to the Khufu pyramid, which was most probably entirely covered with limestone blocks, the Khafre pyramid's base layers were cased with blocks of various granite types from the large quarry area south of Aswan. A more precise localization of the remaining granite casing blocks to defined quarries in the Aswan area is possible with the Aswan map of KLEMM/KLEMM[104], which is based on structure and colour differentiation. According to this map, block sizes similar to the Khafre casing stones were obtained at least until the New Kingdom exclusively from large granite boulders that covered the entire granite exposure there. Consequently, all of the different granite types of that area were incorporated into the various buildings of the Khafre complex. This can be best observed in the Khafre valley temple, where the internal structures offer an almost complete spectrum of the granite types of the Aswan granite plain. Expensive investigations for an exact localization of the granite casing and temple blocks in the Aswan area would only result in a more or less statistical distribution pattern of the entire plain, but such a result might be also achieved by a simple macroscopic correlation of samples with the respective colour plates in the appendix of KLEMM/KLEMM[105].

Granite quarry work was performed rather primitively until the New Kingdom. After the selection of a suitably loose boulder, only the weathered outer coat had to be knocked off. This was done with 10–15 cm large dolerite hammers. Using the same method, the required block size was then roughly cut down to size and transported by

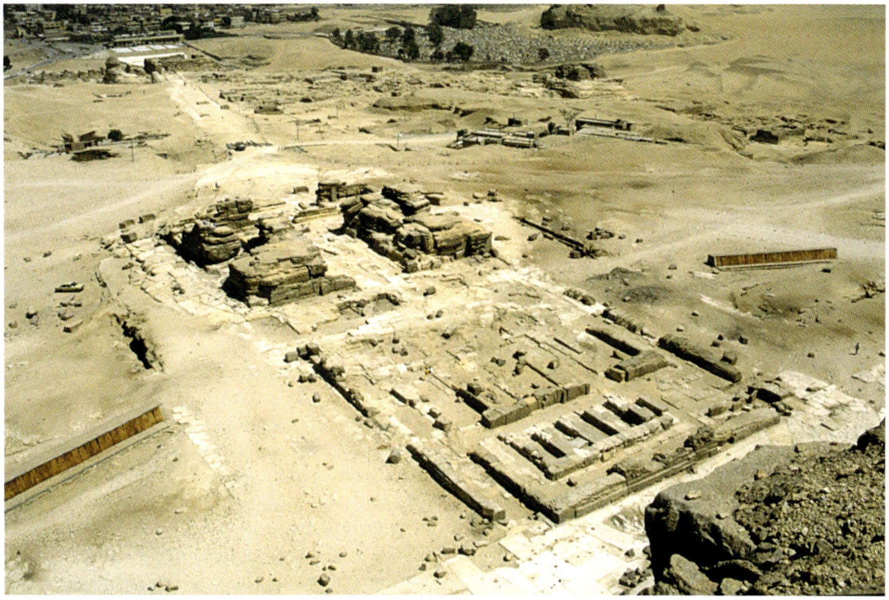

Fig. 98: Mortuary temple of Khafre mainly constructed of local limestone and mud bricks (back). It is cased with fine Tura limestone (foreground) and in inner parts with granite from Aswan.

boat to the building site, where they were finally dressed. An example of this can be seen at the base of the Menkaure pyramid (see Fig. 105).

However, it should be emphasized that only the coarse-grained porphyric granite types were selected for Khafre's buildings. Other granitoid stone types that also occur in the Aswan region, such as granodiorite or the fine- to middle-grained Koror-granite were avoided.

Geochemical provenance determination of the core material

The core masonry samples from the Khafre pyramid correspond well with the samples of the various quarries in the Gizeh area (Figs. 99–101). It further becomes evident that the majority of the core masonry used for the Khafre pyramid is almost identical to the masonry of the Khufu pyramid and must, therefore, have derived from the same sources. This is especially the case for the large quarry area, which encompasses the area from the asphalt road between the Sphinx and the Khufu pyramid (*Route Touristique*) in the north to the Menkaure causeway in the south. But it seems that during the construction of Khafre's pyramid the majority of the stone material was procured from the more eastern parts of the main quarry area known today as the "Central Field". It was used after Khafre's reign in the 5th and 6th dynasties as a necropolis. In this case, it is logical that the later causeway was used originally as a drag ramp that ran directly into the eastern part of the quarry area (see Fig. 73).

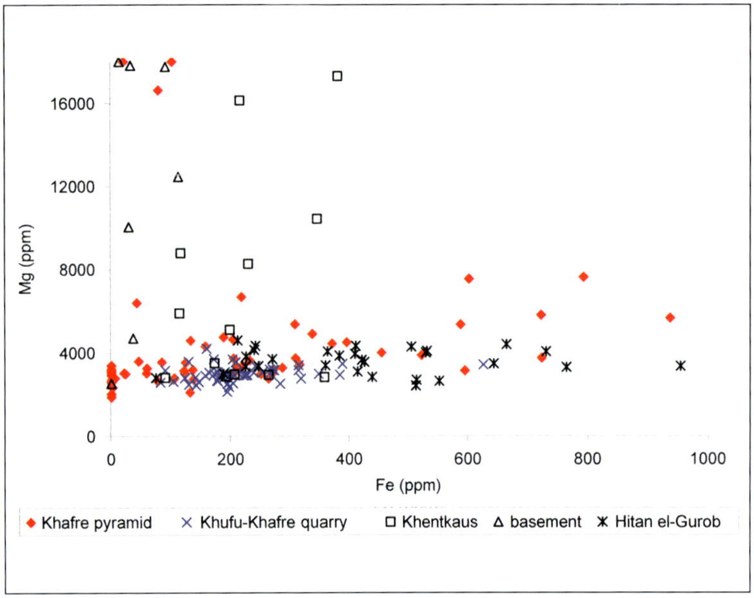

Fig. 99: Mg/Fe-diagram of Khafre pyramid core and the assumed quarries. Most of the samples plot into the main quarry sites east of the pyramid, the pyramid basement and to some extent Hitan el-Gurob.

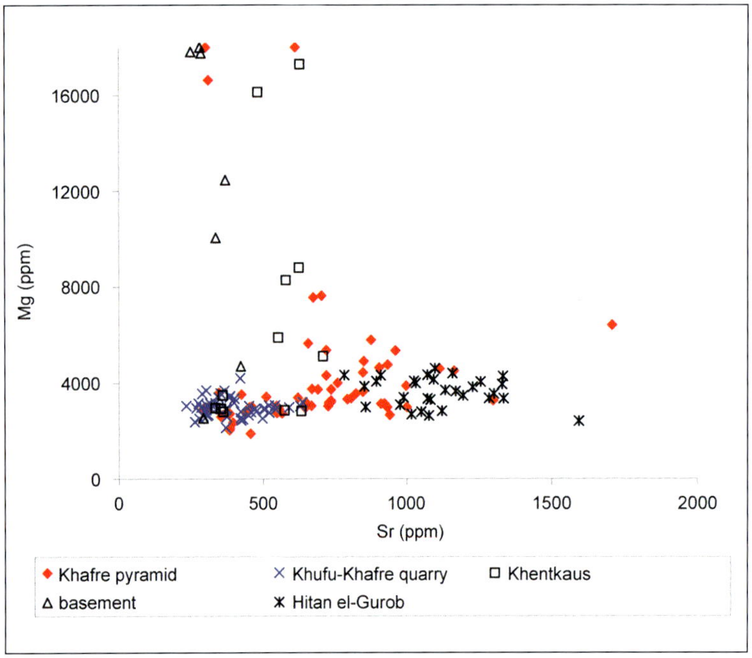

Fig. 100: Mg/Sr-diagram of Khafre pyramid core and the assumed quarries. Again, most of the samples plot into the main quarry sites, like those of Khufu and Khentkaus, the pyramid basement and to some extent Hitan el-Gurob.

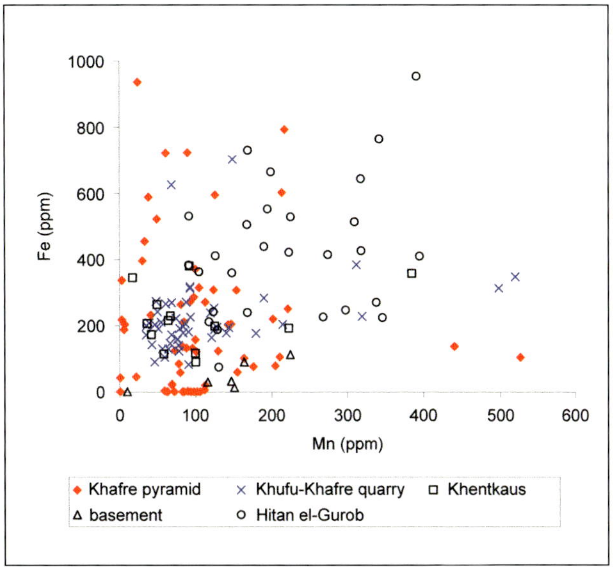

Fig. 101: Fe/Mn-diagram of Khafre pyramid core and the assumed quarries. A good coincidence of most of the pyramid samples with the main quarry sites at the Gizeh plateau is obvious.

Geochemical and petrographical investigations do not lead to a differentiation between the Khufu and the Khafre quarry areas, which both occur more or less in the same geological horizon of dolomitic and marly limestones of the Observatory formation[106].

Finally, the geochemical data indicate that at least a part of the building stones derived from the hill group of Hitan el-Gurob, south-east of the pyramid, and south of the present day Islamic cemetery.

Geochemical provenance determination of the casing and backing material

At least 90 % of the Khafre pyramid was originally covered with a fine limestone quality. Only a fraction of the original casing remains on the uppermost parts of the building. Fragments of the original casing can still be found in the surrounding area; especially the remains of backing blocks allow sufficient sampling of the fine uniform white-grey limestone quality that was used for both casing and backing.

The correlation diagrams Figs. 102–103 show a confined distribution field for the analysed samples. Comparing it with the field of the Tura-Maasara samples, the localization to a well-defined part of this quarry area is unquestionable. An attribution to the Mokattam area may also be excluded due to the micro-facies of the rock structure.

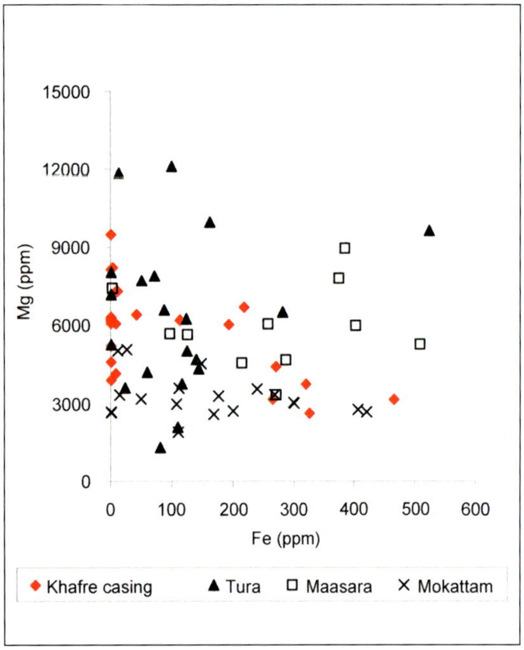

Fig. 102: Mg/Fe-diagram of Khafre pyramid casing and quarries of Tura and Maasara. It turns out that the Tura quarries are the more likely sources.

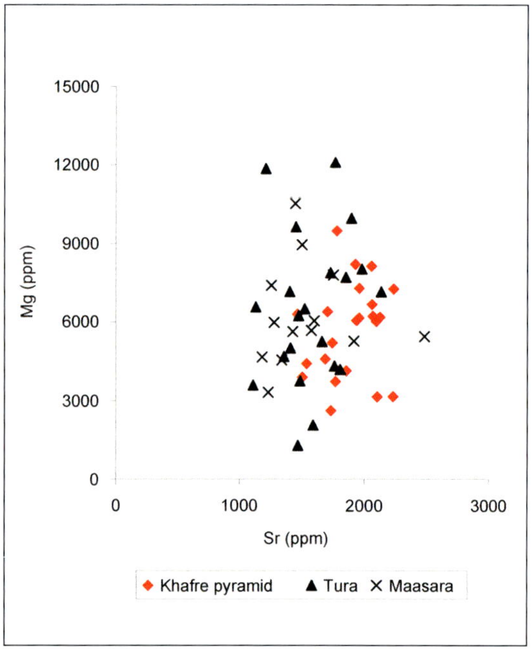

Fig. 103: Mg/Sr-diagram of Khafre pyramid casing and quarries of Tura and Maasara. Again, the close affinity to the Tura sites is obvious.

3. The Pyramid of Menkaure (2539–2511 BC)

Geographical position (present entrance): 29°58'22" N, 31°07'41" E

Original height: 66.45 m | Base: 104.6 m | Slope: 51° | Length of causeway: 608 m

The Menkaure pyramid is built at the far end of the Gizeh diagonal on the edge of the Mokattam formation (Fig. 104), where it dips down to the south and disappears into the younger Maadi formation (see Fig. 73). The complex includes a valley temple, a causeway, and a mortuary temple on the east side of the pyramid.

The pyramid's base lies 2.5 m higher than Khafre's pyramid and occupies only a quarter of the area used by the Khafre and Khufu pyramids. With its original height of 66 m, Menkaure's pyramid represents only about a tenth of the mass in comparison to the Khufu pyramid.

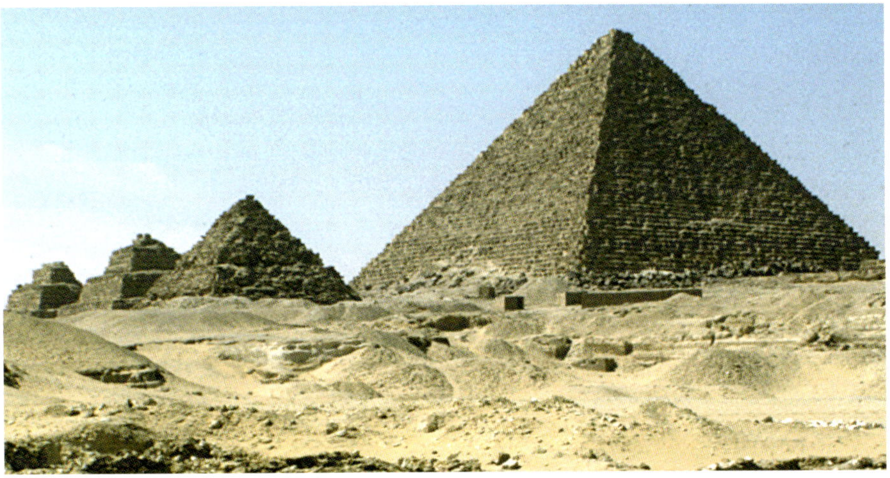

Fig. 104: Menkaure pyramid with subsidiary pyramids, view from south-east.

Again, the core consists of local limestone blocks. Approx. the bottommost 15 m of the pyramid were cased with granite blocks from Aswan. Further up, the casing was made of fine limestone. The final dressing of the granite casing was not completed until the end of the construction process (Fig. 105).

An entrance on the north side provided original access to the inner chambers, about 4 m above ground level. An inclined passage, more than 30 m long, leads to the subterranean chambers. One of them, the proper burial chamber, was constructed entirely with granite, and housed a dark grey sarcophagus, which was found to be empty. After its discovery, the sarcophagus was transported to Europe by ship but was lost when the ship sunk.

The valley temple was built predominantly in mud brick. Some parts of the pavement and column bases were made of limestone. In the inner chambers, the famous greywacke and siltstone triads and fragments of other statues of Menkaure were found.

The causeway leads into the mortuary temple on its eastern side and directly into the central court.

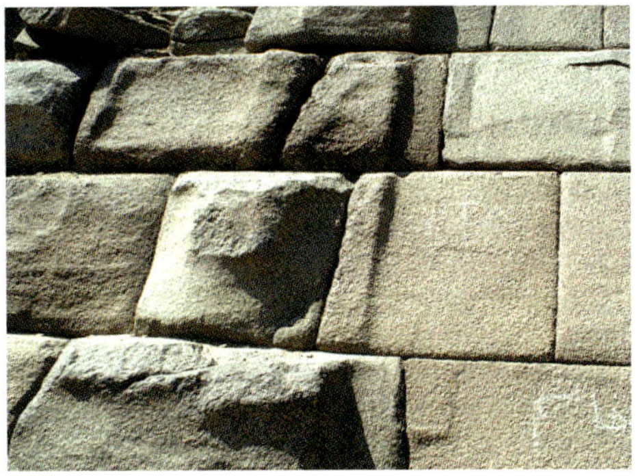

Fig. 105: Detail of Menkaure pyramid. Granite casing blocks at its northern site in various states of smoothing.

The whole building gives the impression that it was finished in haste, most probably by Menkaure's successor Shepseskaf, who made extensive use of mud bricks instead of stone masonry. Some parts were cased with granite and granodiorite. Limestone was used as casing in only a few cases.

Three small pyramids stand to the south of Menkaure's monument. The eastern one was probably the true subsidiary pyramid. It was partly cased with red granite, and a sarcophagus of the same material was sunk into the floor of the burial chamber.

REISNER[107] excavated the entire complex between 1906 and 1924. Recent excavations around the pyramid of Menkaure have been conducted by the SCA.

Geochemical provenance determination of the building material

The geochemical correlation diagrams of the Menkaure pyramid's core material unequivocally assign the stones to the quarry area, situated south-east of the pyramid (see Fig. 73).

Of course, this is not really new information: As rock tombs contemporary to Menkaure were cut into the quarry walls, this site was always regarded to be his quarry[108]. This can now be confirmed because of the good geochemical correlation between the quarry samples and the samples from the core blocks (Figs. 106–109). This result corresponds well with petrographical observations and confirms a former attribution resulting from a smaller database[109].

The limited quarry area to the south-east of the pyramid (Fig. 79c) belongs stratigraphically to the same unit as the great quarry area used by Khufu and Khafre but lies at a higher level. Thus, slight petrographical and geochemical variations from these were expected. Comparing the geochemical data from the Menkaure quarry with data from the more western parts of the central quarry, a significant difference is not really recognizable.

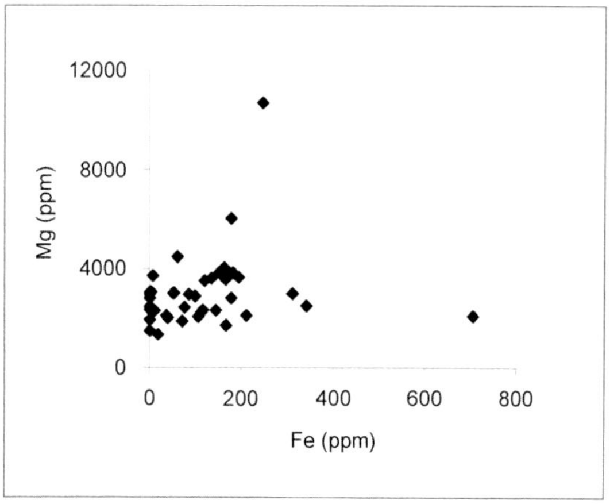

Fig. 106: Mg/Fe-diagram of Menkaure pyramid core material.

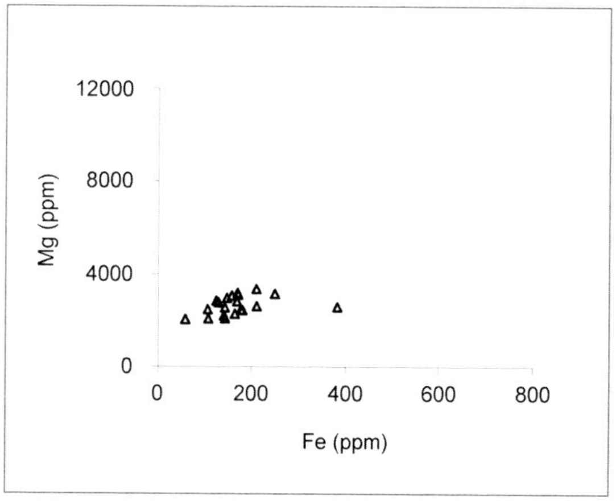

Fig. 107: Mg/Fe-diagram of Menkaure quarry material. Both core and quarry limestones plot in almost identical distribution fields.

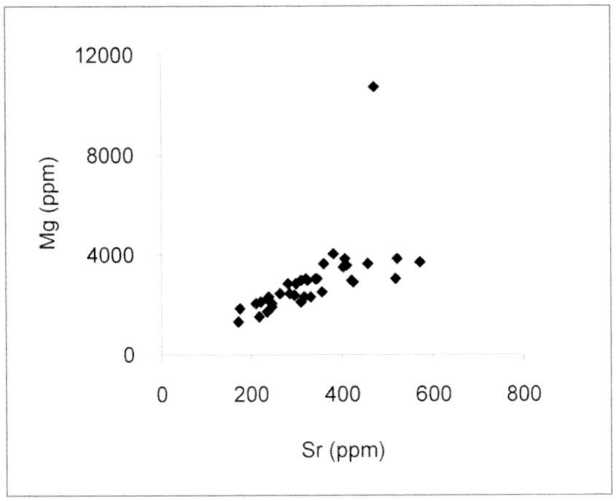

Fig. 108: Mg/Sr-diagram of Menkaure pyramid core material.

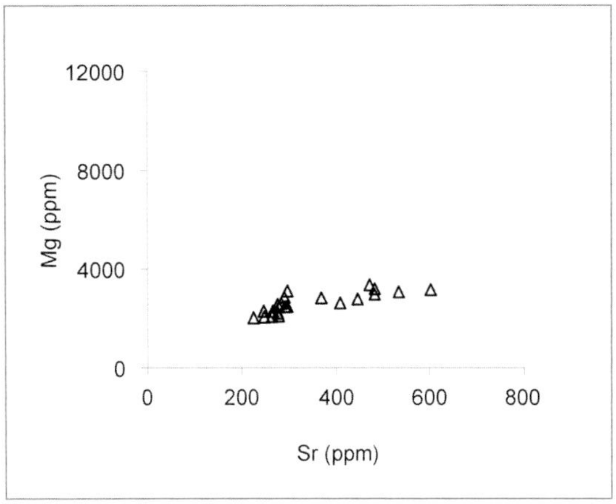

Fig. 109: Mg/Sr-diagram of Menkaure quarry material. Both core and quarry limestones plot in almost identical distribution fields.

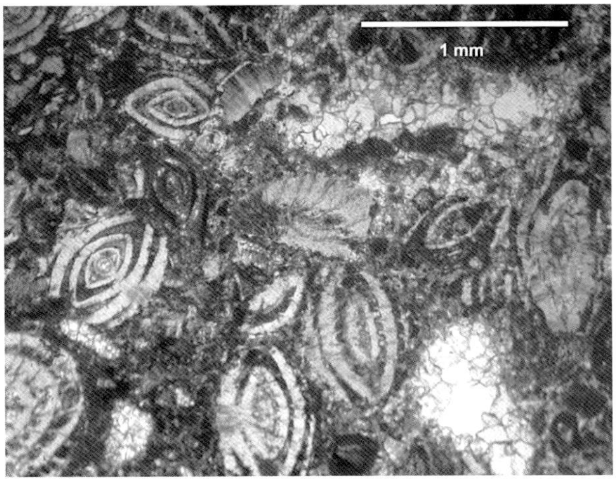

Fig. 110: Micro-photo (QS 1361) of an average thin section of Menkaure quarry limestone with Nummulites gisehensis (1 pol. filter).

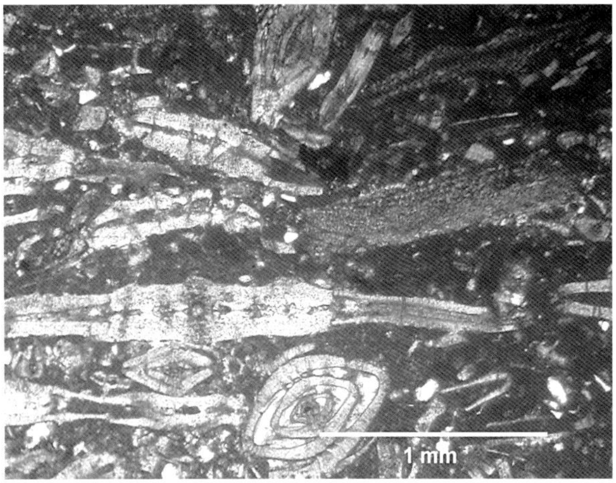

Fig. 111: Micro-photo (QS 1356) of an average sample of Menkaure quarry limestone with Nummulites gisehensis and discocyclinae (1 pol. filter).

Macroscopically, the quarry limestone positively correlates to the pyramid's core material, and it is hard to differentiate it from the Khufu quarry in the western part of the great quarry area.

Under microscope, studying thin sections of both the Menkaure quarry and the core masonry, a somewhat higher porosity can be observed, which is due to the many open fossil interspaces (Fig. 110). Further, a denser structure of the

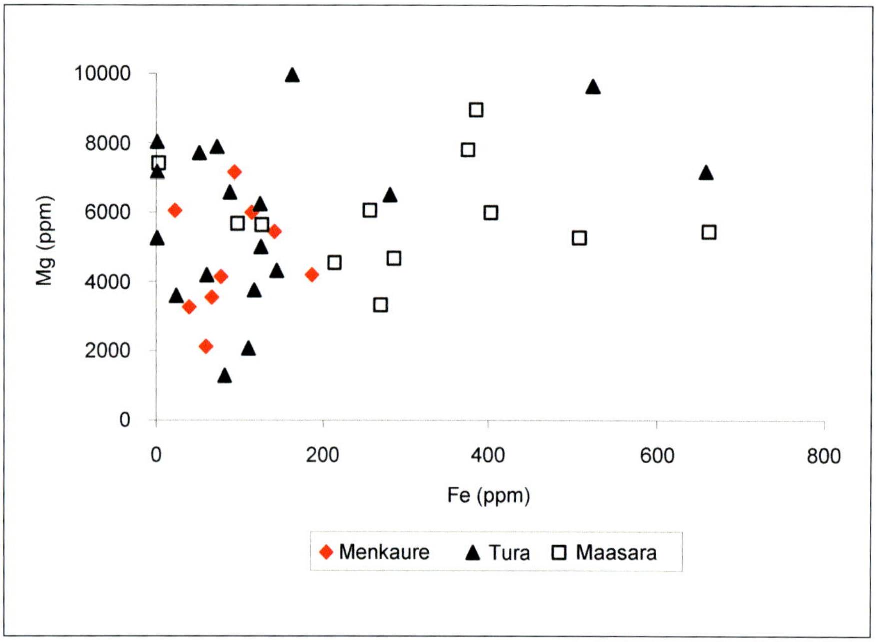

Fig. 112: Mg/Fe-diagram of Menkaure pyramid casing material and limestones from Tura and Maasara. The closer affinity to the Tura samples is conspicuous.

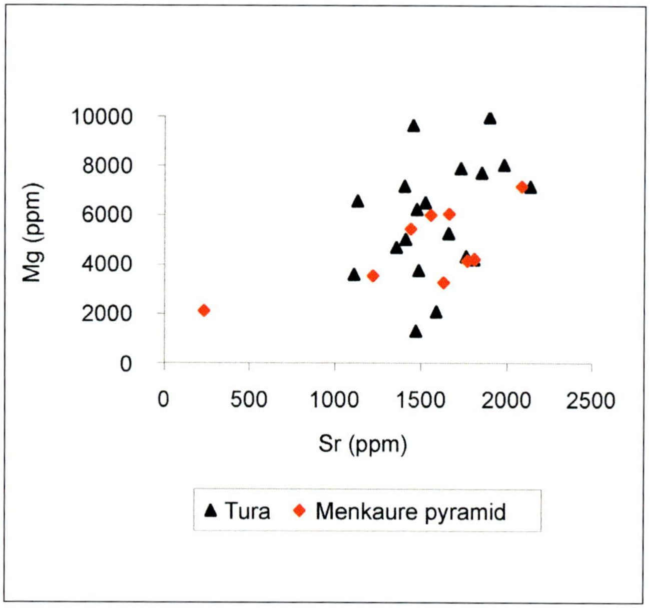

Fig. 113: Mg/Sr-diagram of Menkaure pyramid core material and Tura limestones. Even the only low sample population plots closely together with the Tura field.

many small nummulites, discocyclinae, and other fossil remains is evident (Fig. 111). Generally, a differentiation from the other quarry sites of the Gizeh plateau seems unrealistic.

The casing material of the Menkaure pyramid is again of bimodal composition: Remains of apparently *in situ* granite casing blocks indicate that at least the lower 16 stone layers of the monument were covered with Aswan granite, which roughly corresponds to the first third of the lateral face. On the northern face, the granite casing is still intact until the 7th layer. Similar to the Khafre complex, granite blocks of various qualities of the Aswan granite were incorporated into the casing, but in the case of the Menkaure pyramid it seems that more homogeneity of colour was desired and that pinkish red rock qualities were preferred.

In addition, fine limestone was also used as casing material, but it remains only as backing stones. The use of this limestone is proved *in situ* at the pyramid but also by individual blocks that can be found around the pyramid. The original height of the limestone casing can no longer be determined. In the frame of the present sampling program, the main emphasis was laid on the backing blocks that remain in their original position rather than on the individual loose blocks, which were sampled only for control.

Once more, the geochemical data plotted in the correlation diagrams of Figs. 112 and 113 show close concordance. A comparison of the casing and backing limestone values with the Tura limestone data allows the conclusion that Menkaure extracted these stones from a very limited area of the Tura quarry region, most probably from one distinct gallery quarry. In Fig. 112 the Maasara quarry samples correlate poorly with the Menkaure casing and backing stones, to such an extent that this area can be excluded as a possible material source. Further, comparing data with the corresponding field of Khafre it is clear that the fine quality building material extracted in Tura derived from different quarries there, a procedure that is repeated for the later Old Kingdom pyramids.

VII. The Pyramid of Djedefre (2581–2572 BC) at Abu Roash

Geographical position: 30°01'57" N, 31°04'31" E

Estimated height: ~ 65 m | Base: 106 m | Slope: 52° | Length of causeway: 1.5 km

Djedefre, the 3rd ruler of the 4th dynasty and the son of Khufu, abandoned the necropolis at Gizeh and built his pyramid at Abu Roash, 4 km north of the Gizeh plateau. It is the northernmost royal pyramid in Egypt, but it was never completed. His successor Khafre then returned to Gizeh.

The first systematic investigation of the pyramid was undertaken in the middle of the 20th century[110]. After a first survey of the site by MARAGIOGLIO/RINALDI[111], a French-Swiss team began excavation there in 1995[112].

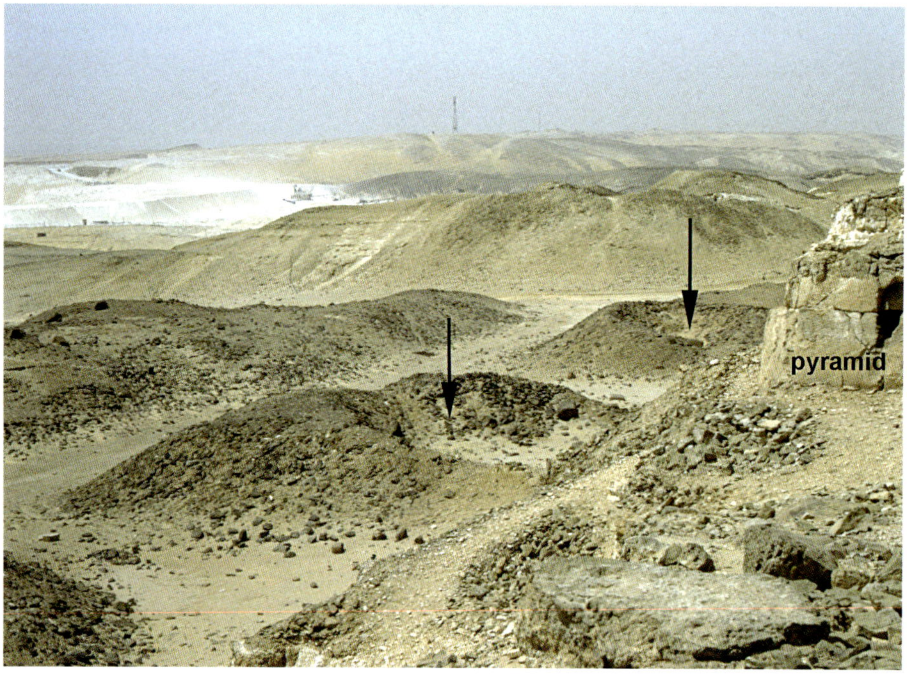

Fig. 114: Djedefre pyramid: Working sites for Aswan granite blocks at the west side of the pyramid (arrows; photo taken 1987).

The causeway runs unusually from north to south and is about 1.5 km long. For the most part, it encompasses the natural slope of a distinct limestone layer and meets the enclosure wall to the north of the pyramid in a large open space.

Only little remains of the valley temple at the foot of the causeway located at the mouth of Wadi Qaren[113].

Recent excavations have unearthed an inner enclosure wall about 6 m from the pyramid's northern side that widens in the east, where mud brick remains are probably the remnants of a mortuary temple, which was apparently completed in haste. This mortuary temple was formed of rather thick stonewalls, finished with mud brick to form the compartments and chambers that surround an open courtyard.

In the north-east corner of the inner enclosure, workshops and housing have been found that were used by the builders of the pyramid. Layers of stone chips remain from what was apparently a stone yard where the pyramid blocks were worked. Along the western and southern sides, several open workshops with remains of granite working were discovered (Fig. 114), but only one is existent on the southern side at the south-eastern edge.

As is the case with a number of pyramids in Egypt, outcropping rock was used as part of the core. Today, the core consists of local limestone blocks in approx. 15 horizontal layers. It was obviously planned to cover the pyramid to some extent with granite, but very little of the red granite casing still remains.

Though the substructure is highly damaged, we can see that the builders used the open pit method of construction. In the north wall of the pyramid, a trench was built that now contains the remains of a descending corridor. The corridor generally follows a north-south axis, and leads to two rooms, an antechamber, and a burial chamber. A fragment of a red granite sarcophagus was found[114].

On the eastern side of the pyramid, fragments of some 120 statues were discovered within the enclosure wall in a trench similar to a boat pit. The statues mostly represent Djedefre enthroned, among them the masterpiece of the king's head (Louvre). They were made of red quartzite from the quarries of Gebel el-Ahmar, near ancient Heliopolis.

In addition, a subsidiary pyramid without a burial chamber stands near the south-east corner of Djedefre's pyramid.

Geological environment of the pyramid

Djedefre's choice of building site about 4 km north of the Gizeh plateau resulted in an even more exposed hillside position than that of the Khufu pyramid. When investigating this site one feels instinctively sorry that this pyramid remained unfinished, as it would have been of comparative importance as the Gizeh complex if ever completed.

The geological base of the pyramid is older than at Gizeh and dates to the Upper Cretaceous. The intensive folded limestone units west of Cairo to the Alexandria desert road are conspicuous when passing the hill range about 7 km behind the Gizeh road junction (Fig. 115).

This is the result of an intensive tectonic folding phase and a subsequent faulting at the end of the Cretaceous period. During the later Tertiary and especially Eocene limestone units, such folding only took place south of Luxor.

Appropriate to the impressive geology, this area was intensively investigated geologically. It would go beyond the scope of this presentation to only mention the most important contributions, all the more as the Cretaceous limestones had, astonishingly enough, no significant impact on the ancient Egyptian building stone record. The geological sequences of the Upper Cretaceous in this area include Cenoman, a formation with sandstones and fossil-rich clay intercalations (the Bahariya formation according to the Geological Map of Greater Cairo Area[115], followed

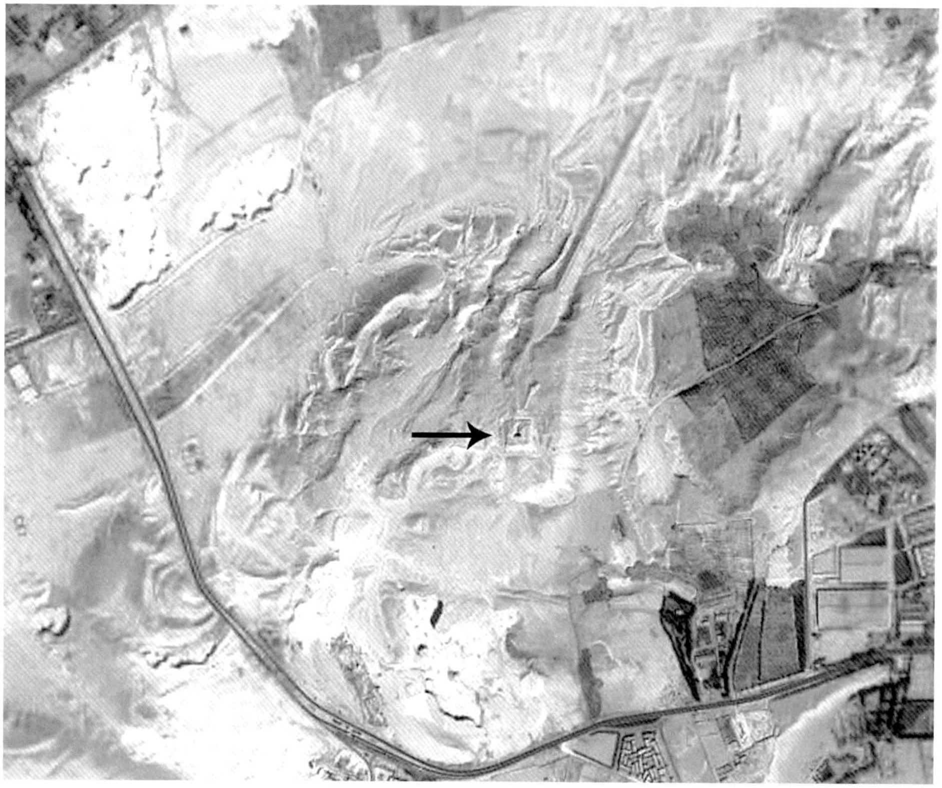

Fig. 115: TM satellite photo (TM scene 176/39, band 7) of the Abu Roash area with intensive folding of Cretaceous dolomite- and limestone formations (arrow: Djedefre pyramid with causeway leading north-east).

by Turonian limestones, marls, and dolomites with flint concretions and basal clay horizons. These Turonian limestones belong to the so-called Wata formation, can be up to 130 m thick and form the hills in this area together with the base of Djedefre's pyramid, including the core masonry. The stratigraphically following fossil-rich limestones and marls of the Santonian and Maastrichtian are of no significance for the present study.

Of importance, however, is the flat inclination of the Wata formation, which extends towards the north-east into the wide Wadi Qaren that runs from the south-west 1 m above the fertile plain. With a length of approx. 1.5 km, the longest causeway of all the Old Kingdom pyramids follows this natural flat inclination of the fertile plain (see Fig. 115). At the end of the causeway in Wadi Qaren, a harbour could be easily connected with the valley temple due to the topographical similarities between the cultivated land and channel system. Only little remains of the original harbour[116].

At the proposed construction site, a flat hill originally existed. The burial chamber and sloping passage were excavated into the more or less horizontal limestone layers without any major difficulties. The stone blocks extracted during this process were used for the construction of the pyramid's base. In addition, an easily accessible limestone hill (occupied today by a radar station) situated a few hundred metres away from the pyramid's position was an ideal supply area for good quality rock material. Thus, the geological framework for the construction of a large pyramid at this site was optimal.

Provenance determination of the core material

In spite of intensive search for the remains of a quarry near the pyramid, nothing of the kind could be determined with any certainty. Obviously, the small amount of rock material for the construction of only the base layers did not leave significant traces with the exception of the burial chamber and the layout of the wide shaft.

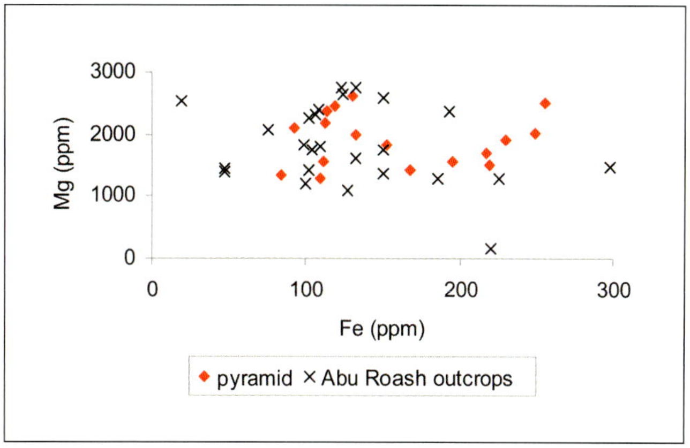

Fig. 116: Mg/Fe-diagram of Djedefre pyramid socle in comparison with the outcrops from its direct environment. A relatively good coincidence of both occurrences is evident.

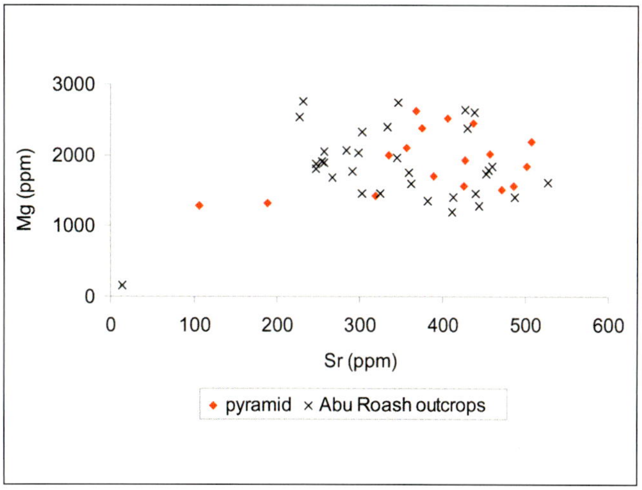

Fig. 117: Mg/Sr-diagram of Djedefre pyramid socle in comparison with the outcrops from its environment. Again, a relatively good coincidence is evident.

After identifying the macroscopic identity of the core masonry and the outcropping rocks of the inclined shaft, the burial chamber, and the pyramid's closer environment, the decision was made to statistically sample the pyramid's surroundings in order to clarify the precise provenance of the core masonry rocks. An emphasis was laid on the hilly outcrops extending west of the mastaba field and south of the Abu Roash village. Unfortunately, clear structures that would indicate an ancient quarry were not found. Recently, however, the actual quarry site was found in the area: it lies 1.8 km north-east of the pyramid complex in the hill outcroppings of Gebel Madawarah, south of the Abu Roash village. At least 200 000 m^3 of limestone were excavated there[117]. Thus, this area fits well with previous sampling.

In the correlation diagrams of Figs. 116–117, it is evident that most of the plotted values cluster within a very limited field, which indicates that the entire material comes from a limited area with only a slight geochemical variation. A comparison with the plotted values of the outcrop samples shows that the right rock unit of the core masonry was sampled, even when not directly in the actual quarry.

The provenance of the granite casing remains to the Aswan area is completely undoubted. Some of these casing blocks remain *in situ* in the pyramid's base and are held together by a thick layer of mortar to the limestone core masonry. The mortar itself contains fragments of granite aggregates.

In some of the well-recognizable working places around the pyramid's base, different stages of stonemasons' treatments of the granite blocks that had been delivered in a coarse state to final casing blocks could be observed. When analysing the chips in the waste heaps of these working places, it is clear that the chips occur in very different states of deterioration. This allows the conclusion that the granite blocks were delivered as blanks. Once at the site, their weathered crust was cut off and the block shaped into the requested form. This was done with dolerite hammers, which were also found in great numbers, sometimes still sharp-edged, broken, or already rounded, in the waste heaps.

The casing blocks of granite that still remain today had already been smoothed, and it may be assumed that they were incorporated in this state into the pyramid's base, in contrast to the younger Menkaure pyramid.

Quite a number of the remaining granite blocks in the working places show traces of wedge splitting. This technique was introduced by the Romans but continued until Byzantine and even Arab times. STADELMANN[118] assumes that stone robbery at the site dates back to New Kingdom times, but no traces could be detected during the present study to confirm this hypothesis. The extent of stone robbery could not have been of vast dimensions as granite blocks remained in some of the original working places at least until these sites were recently removed/excavated at the southern side of the pyramid.

VIII. The Mastabat el-Faraoun of Shepseskaf (2511–2506 BC)

Geographical positions: Mastabat el-Faraoun: 29°50'21" N, 31°12'54" E
Quarry: 29°49'26" N, 31°11'20" E

Height: 18.9 m | Base: 99.6 × 74.4 m | Slope: ~ 70°

Unlike his immediate predecessors and his successors, Shepseskaf chose a site at Saqqara-South for his tomb (Fig. 118). Probably, there was no suitable space anymore on the Gizeh plateau to erect a comparable pyramid as his ancestors did. Called by the locals "Mastabat el-Faraoun", it takes the shape of a huge royal sarcophagus.

In all certainty, there was a valley temple connected with this tomb, but it has never been excavated. The causeway led to the south-east corner of the mortuary temple, which is adjacent to the eastern side of the building, before it is running along the south wall of it into an open courtyard. The temple walls are built from mud brick, but its inner parts are paved with limestone. One pilaster in red granite still stands upright, however, the majority of granite and granodiorite blocks are scattered around the site[119].

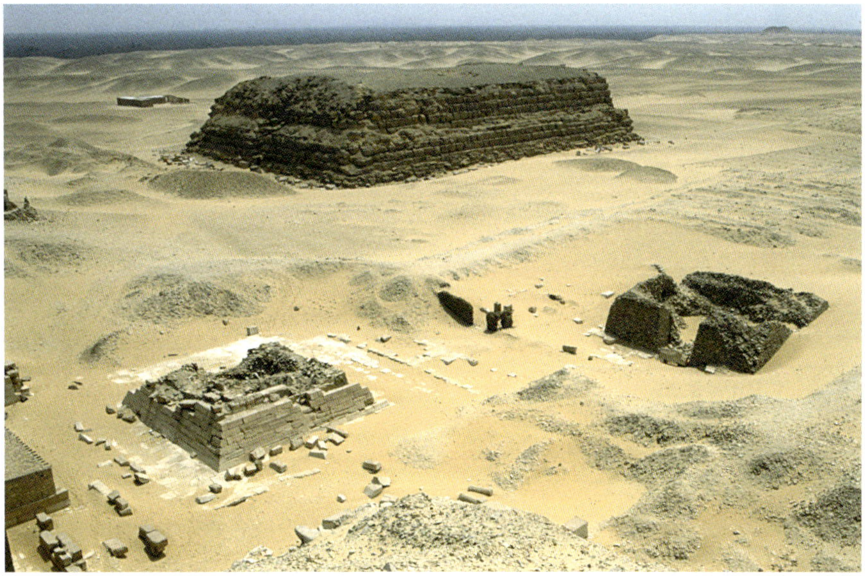

Fig. 118: Mastabat el-Faraoun. View from north-west on top of Pepi I's pyramid with its cult- and Udjebten (right) pyramids in the foreground.

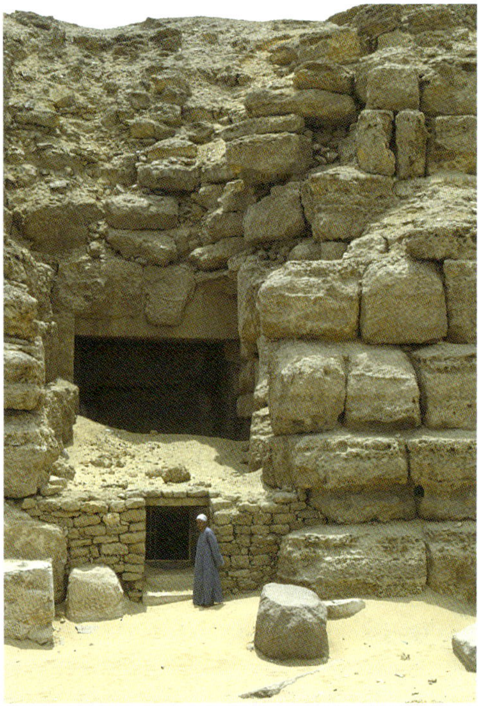

Fig. 119: Very large block sizes at the northern entrance of Mastabat el-Faraoun.

The core of the mastaba was built in two levels using large rectangular blocks of local limestone measuring 2.5 × 1.5 × 1 m in average size (Fig. 119). The building was cased with fine limestone except for the basal course of red granite. On some of the casing blocks, inscriptions of Khaemwese's later restoration were found[120]. On the axis of the north wall, about 2.5 m above ground level lies the entrance to the substructure. It leads to an antechamber, which gives access to the burial chamber constructed with a red granite ceiling. The remains of a broken sarcophagus of greywacke are still *in situ*; however, the mastaba was probably never used for a burial. The Mastabat el-Faraoun measures 99.6 × 74.4 m and is oriented north-south.

Core material

Unfortunately, the quarry site of the Mastabat el-Faraoun could not be inspected as it is situated in an off-limits military area, but its localization is well indicated by a large drag ramp that runs 2.3 km south-west from the mastaba's south-west corner, only intersected after approx. 700 m by a modern railway track (Fig. 120). The relicts of the quarry area are still well recognizable from a distance. Because of the military restriction, this limestone material could only be studied at the Mastabat el-Faraoun itself.

Macroscopically, the rock type is a relatively porous, brown to reddish-brown calcareous sandstone to sandy shell-rich limestone, which is rich in fossil debris and also oyster shells in parts with mother-of-pearl remains

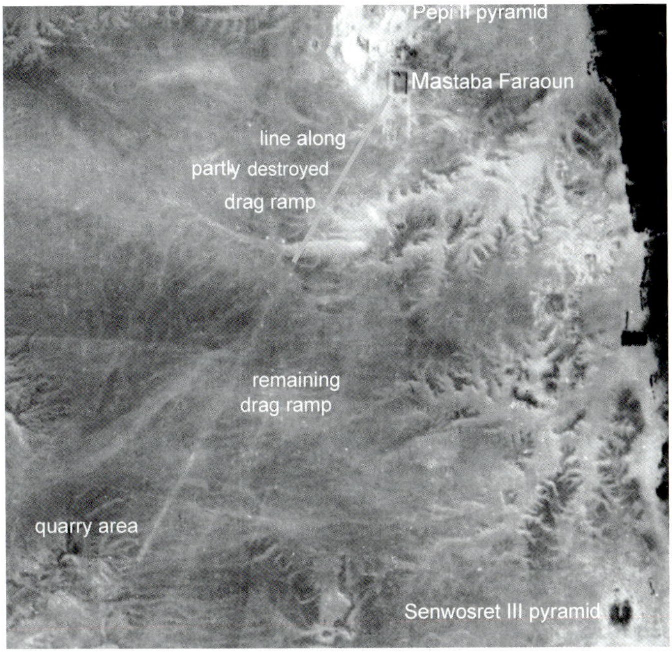

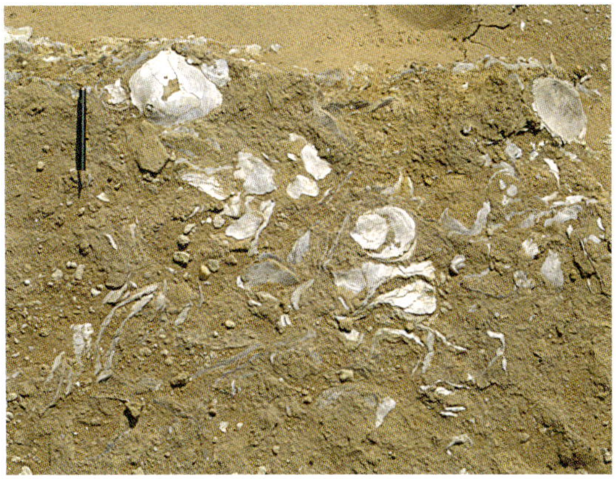

Fig. 120a: Sandy limestone from Mastabat el-Faraoun with oyster shells (16 cm pencil for scale).

◀ *Fig. 120: Long drag ramp (marked) from Mastabat el-Faraoun into its quarry area (detail from aerial photograph P 14, Faiyum 3–49, 1954).*

(Fig. 120a). The coarse, porous character and the large oyster shells up to 6 cm in size along with the remains of mother-of-pearl are the most striking features.

The same quarry material is more easily accessible at the quarry area of the Red Pyramid. Here, two drag ramps run from the pyramid's south-west corner into the heavily sand blown quarry areas at the edge of a slightly raised limestone bank. The limestone here is again very porous, but contains less fossil debris and rarely oyster remains. A more detailed description is given above in connection with the Red Pyramid. Despite the fact that a relatively low number of samples was taken from both the mastaba's core and the quarry, i.e. restricting representative data, the core masonry can therefore only be correlated with the stone material from the Red Pyramid quarry.

With a hand lens, the fossil debris with its shelly character becomes more distinct and the sandy structures cannot be overseen. Under microscope, this rock type varies greatly. Each section is always dominated by mainly sharp-edged quartz grains, 0. 1–0.05 mm in size, bound by fine crystalline calcite within a disordered structure of mainly shell fragments (Fig. 121), but also echinoderm plates and fish scutes. The Mastabat el-Faraoun samples contain distinctly more fossil debris than those from the Red Pyramid and its quarry areas.

In spite of the lack of original quarry material the geochemical correlation diagrams indicate a good concordance of the Mastabat el-Faraoun core material with that of the Red Pyramid quarry site, which occurs within the same geological unit only some 300 m to the south (Figs. 122–123).

Stratigraphically, these calcareous sandstones to sandy limestones belong to the Kom el-Shellul formation of the Pliocene (upper most Tertiary system), which overlies unconformably the Eocene rocks south of the Gizeh plateau until Beni Suef.

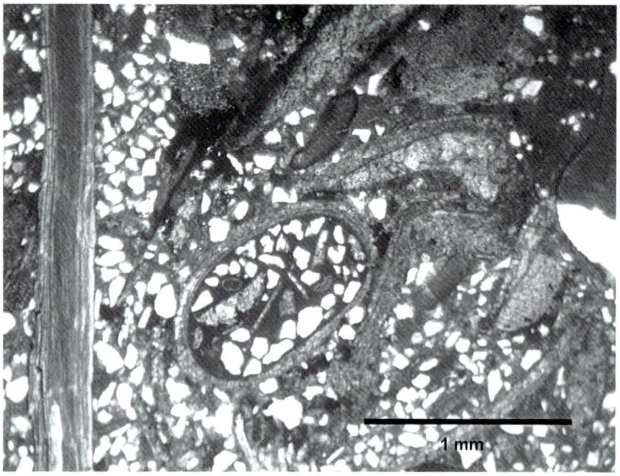

Fig. 121: Micro-photo (Sf 13) of core material from Mastabat el-Faraoun. Sandy limestone (white grains) with a limonite enriched microsparitic matrix and many calcified fossil remains. Note at the left side the long oyster fragment in contrast still in its original aragonitic structure (1 pol. filter).

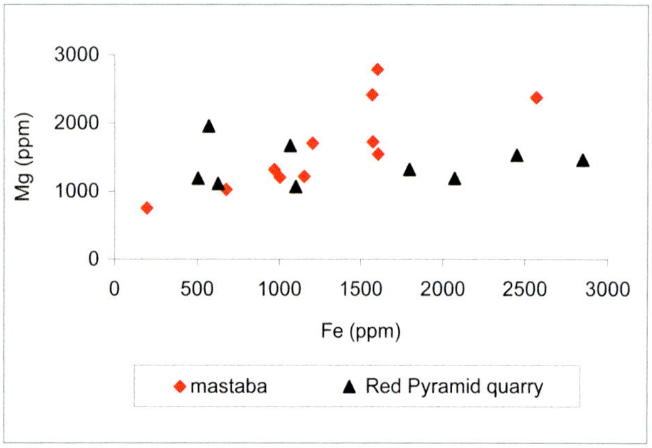

Fig. 122: Mg/Fe-diagram of Mastabat el-Faraoun core and of Red Pyramid quarry. The good geochemical concordance of both depends on the close vicinity of the Red Pyramid and Mastabat el-Faraoun quarry sites, occurring within the same geological unit.

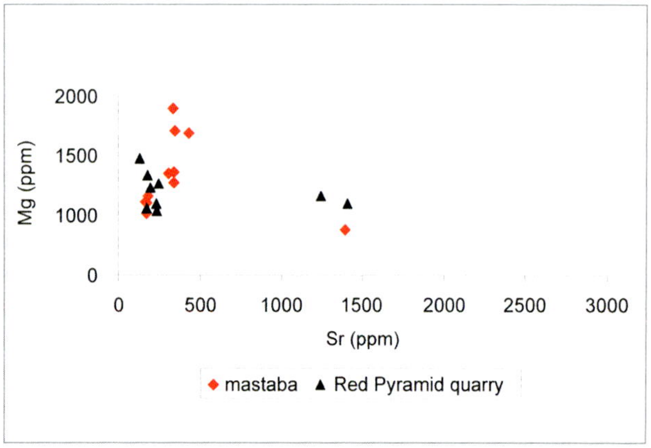

Fig. 123: Mg/Sr-diagram of Mastabat el-Faraoun core and of Red Pyramid quarry. A close geochemical concordance is striking.

Casing material

Only little remains of the white-grey, fine limestone of the casing material. This limestone variety also forms the backing stones and the backing of the burial chamber system as well as the constructions stabilizing parts of the mastaba. An example of the latter can be easily made out above the entrance (see Fig. 119).

Despite few available samples, the casing material can be determined as the Tura type as demonstrated petrographically in Fig. 124.

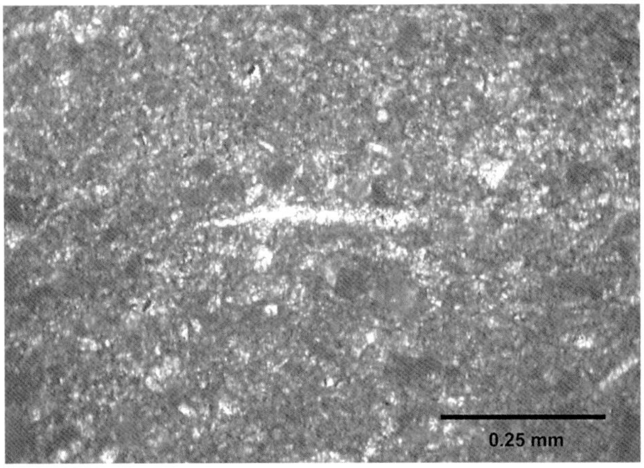

Fig. 124: Micro-photo (Sf 31) of casing limestone from Mastabat el-Faraoun. The typical fine-grained microsparitic matrix of Tura limestone contains rarely sponge needles and very sparsely remains of other small fossils (not in the photo) (1 pol. filter).

The provenance of this fine limestone is clearly indicated by the geochemical correlation diagrams in Figs. 125–126, which allow an attribution to the Tura quarry district. In this case, an even more precise attribution is possible as the data plotted within the limits of error into the field of the large gallery quarry situated next to the Tura cement factory.

Further, the casing of the entrance corridor and the burial chamber consists of huge granite blocks from Aswan, whereby the pink-red variety of the rose granite dominates. Only a few blocks close to the entrance are of granodiorite, again from Aswan, whereas the many granitoid stone fragments around the mastaba are mainly of the darker reddish granite type with parallel texture. Also, fragments of granodiorite from Aswan were found. The definite macroscopic attribution of these granitoid varieties to the quarry area south of Aswan does not necessitate further geochemical and petrographical analysis.

Finally, the Mastabat el-Faraoun itself is based on beige, sandy limestone, which can also be found at the base of the small mortuary temple situated to the east of it. These blocks, the base, and pavement of the mortuary temple derive from the Saqqara-South quarry sites. Thus, even under restricted sampling conditions, well-defined localizations of the various stone qualities used for the construction of this unique monument could be established.

Last but not least, the provenance of the Shepseskaf stone sarcophagus could be petrographically identified as greywacke from Wadi Hammamat instead of basalt, as commonly reported for black sarcophagi.

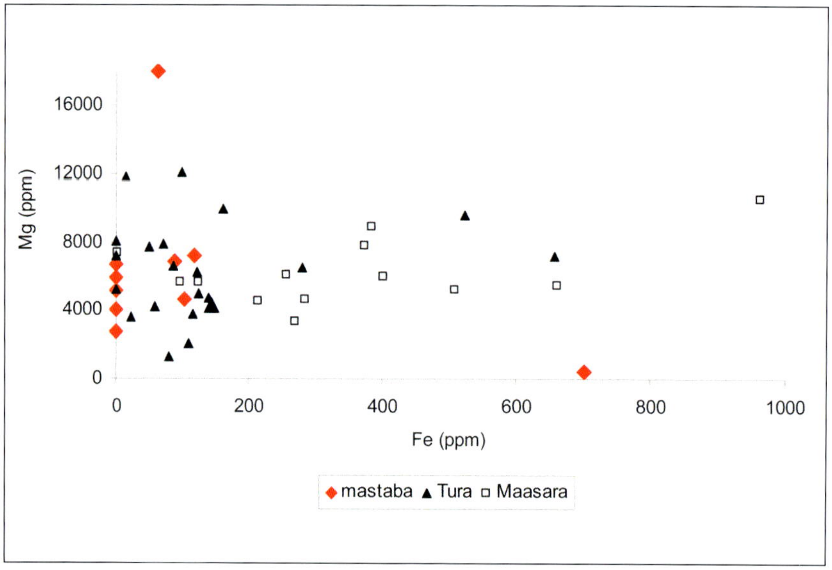

Fig. 125: Mg/Fe-diagram of casing samples from Mastabat el-Faraoun in comparison with quarry samples from Tura and Maasara. The better fitting with the Tura field is obvious.

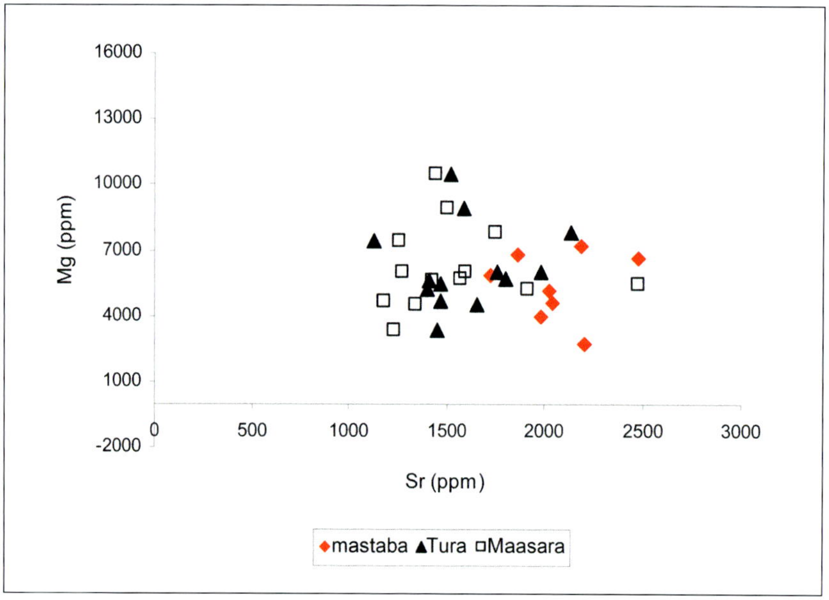

Fig. 126: Mg/Sr-diagram of casing samples from Mastabat el-Faraoun in comparison with quarry samples from Tura and Maasara. The casing samples fit better to Tura quarries.

IX. The Pyramid of Userkaf at Saqqara-North (2504–2496 BC)

Geographical position (entrance): 29 °52'25" N, 31°13'0.7" E

Original height: 49 m | Base: 73.3 m | Slope: 53°

Userkaf was the founder of the 5th dynasty. He built his burial place as a true pyramid in Saqqara-North (Fig. 127), near the north-east corner of the Djoser complex.

The pyramid is built in horizontal layers with relatively large rectangular blocks, bound together by thick mortar fillings. Local sandy limestone was used for the pyramid's core and a fine limestone for the casing. The protective casing only survives in few parts, causing a heavily eroded building today.

The mortuary temple was built at the south side due to topographical reasons; there was not enough space at the eastern side of the pyramid as a wide and deep trench surrounding the whole Djoser complex limited the extent of the site[121]. A large enclosure wall of limestone protected the mortuary temple and the pyramid itself.

The causeway, which leads to the mortuary temple, was originally paved with basalt. The basalt paving can be made out on aerial photographs[122]. A valley temple has not yet been found.

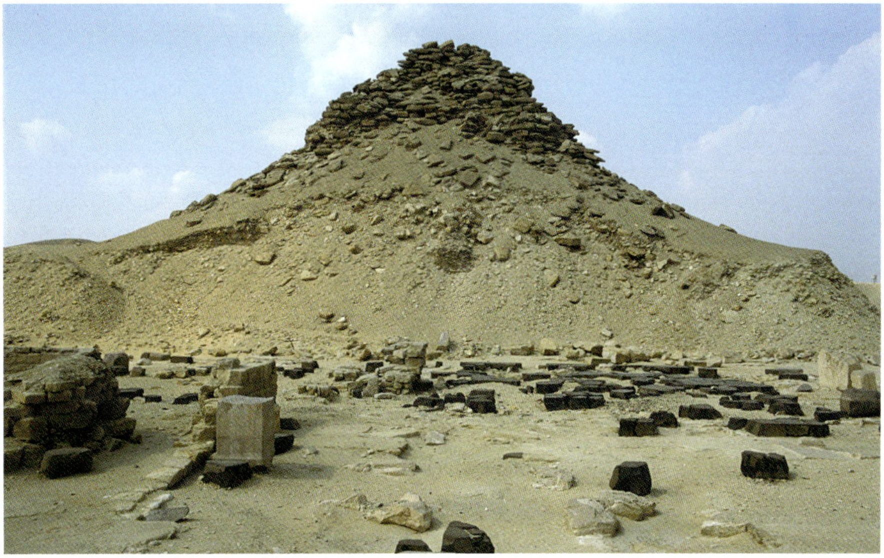

Fig. 127: Userkaf pyramid with mortuary temple and cult pyramid (foreground) south of it.

A small cult chapel was attached to the east side of the pyramid. The sparse remains show a false door of quartzite, a basalt floor paving, and a granite base beneath walls of fine limestone.

The entrance to the pyramid and its underground chambers lies in the north. The corridor leading to the burial chamber was cased with red granite. Fragments of an undecorated black sarcophagus were found, most probably from greywacke as is the case with other black sarcophagi from the Old Kingdom, however, this has yet to be confirmed.

The temple's entrance was situated at the south-east corner where the causeway ended. The most striking feature within the temple is a courtyard with 20 red granite pillars standing on all sides except in the south. The pavement of the courtyard is again of basalt[123].

A small cult pyramid stands at the south-west corner of the pyramid's courtyard. Only a few layers of its core material still remain, but the gape opening into the underground chambers shows a gabled roof, constructed using large blocks of fine limestone.

Petrography

The wide spectrum of sandy to fine-grained marly limestone at the flanks of the escarpment extending towards the Nile valley was already described above in connection with the Step Pyramid's core masonry and the provenance of its building stones. The core material of Userkaf's pyramid shows analogous composition and, therefore, does not need a repeated description here. However, the fine, white-grey casing limestone should be mentioned.

Macroscopically, due to its fine chalky structure and whitish-grey colour, the casing limestone of Userkaf's pyramid can definitely be assigned to the Tura-Maasara district, opposite the western escarpment of the Nile valley.

Under microscope, however, this limestone with its relatively coarse-grained microsparitic matrix with finely milled fossil particles and rare sponge needles shows more of a tendency towards a localization at Maasara, which, however, contrasts with the geochemical results, as demonstrated below.

Geochemical provenance determination

In accordance with the mineralogical and structural bimodality of the core material of Userkaf's pyramid, the geochemical correlation diagrams of Figs. 128–129 also show two distinctly different distribution fields. A comparable bimodality is recognizable in the diagrams of presumable quarry areas, situated mainly along the slope of the escarpment. However, the distribution density of the plotted data is striking. This depends on the predisposition of the more marly layers that were already discussed in the section on the Djoser pyramid to deteriorate. On the other hand, the natural quarry walls have been weathered in parts during the last 4400 years. Consequently, these marly layers were somewhat disregarded during sampling. This deterioration process did not affect the core masonry of Userkaf's pyramid, due to its coverage and therefore protection by the original casing. Thus, likewise the marly building material was sampled in relation to its original existence within the quarried rock unit. Taking this into consideration, the geochemical data show a very good concordance with the quarry samples of the Saqqara-North area. According to Fig. 128, all the sampled quarry sites could be the source of Userkaf's building material. In more general

Fig. 128: *Mg/Fe-diagram of Userkaf pyramid core material in comparison with Saqqara-North quarry sites. A clear bimodality of the pyramids data is obvious.*

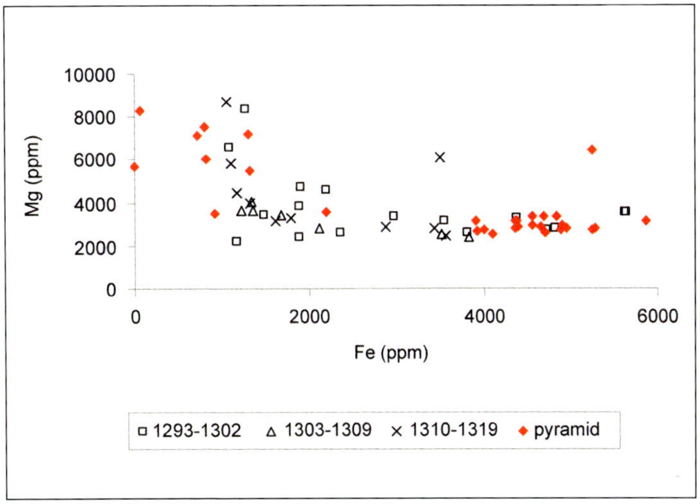

Fig. 129: *Mg/Sr-diagram of Userkaf pyramid core material in comparison with Saqqara-North quarry sites. A bimodality, but no distinct preference to a specific quarry site is recognizable.*

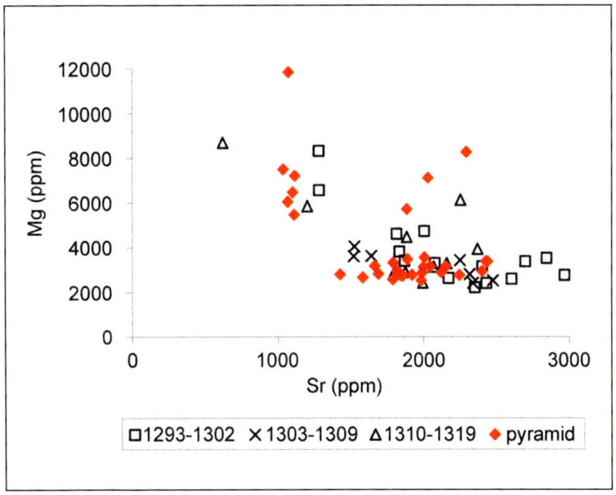

terms, the entire escarpment of the Saqqara-North necropolis is a potential provenance area for this type of core material. These geochemical indications correspond well with the petrographical observations, already discussed in connection with the Djoser pyramid's core material.

In contrast to the monuments discussed so far, the geochemical data of the casing and backing limestone along with the comparable building stones of the satellite pyramid and the temple constructions of Userkaf all point to a new provenance area. As can be seen in Figs. 130–131, the plotted data of the casing samples correspond more closely with the samples from the Tura district than with those from Maasara. This contradicts the petrographical results, and it should be considered that in certain cases the petrography alone might not be sufficient for a specific provenance determination.

At present, a more precise attribution to one of the many gallery quarries at Tura is impossible, mainly due to the fact that sampling was limited to the area of the Tura cement factory. Unfortunately, due to military restrictions, there was no chance to sample the many various gallery quarries within this extended area. This incomplete database, however, does not allow an attribution of the casing stones to a well-defined gallery. Nevertheless, a database that includes all of the various gallery quarries of the Tura area would enable such a precise attribution.

All the granites and granodiorites incorporated into the Userkaf pyramid complex derive exclusively from the quarry area south of Aswan. In the case of Userkaf's pyramid, the light pink-red granite was obviously preferred. The provenance of the basalt material used in the mortuary temple is again the northern Fayum, as is the case with all the Old Kingdom basalts.

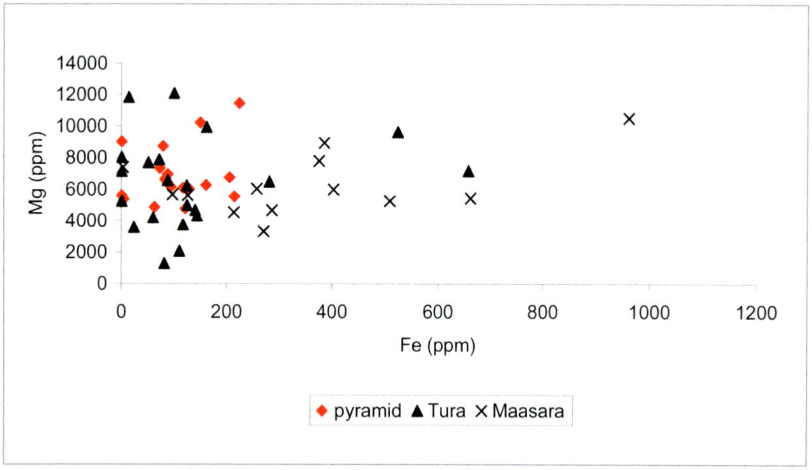

Fig. 130: Mg/Fe-diagram of Userkaf pyramid casing limestone in comparison with the assumed provenance districts of Tura and Maasara. Userkaf casing data indicate a clear preference for the Tura plots.

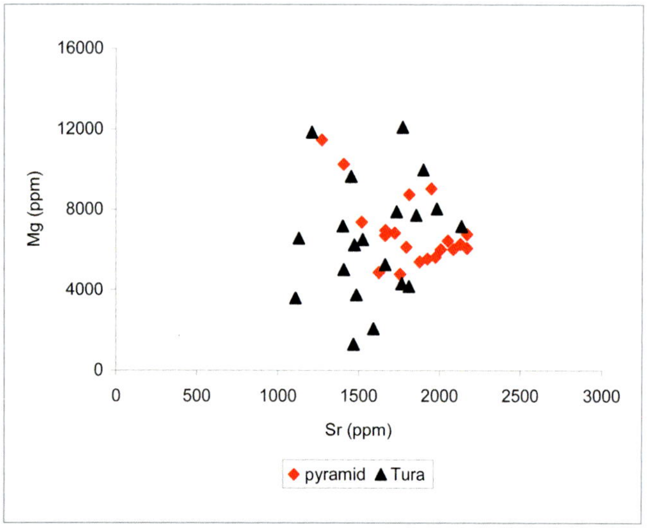

Fig. 131: Mg/Sr-diagram of Userkaf pyramid casing limestone in comparison with the assumed provenance district of only Tura. The data indicate a dense cluster of the casing stones within a rather limited field, which might denote that they derive from an only single (gallery) quarry within the wide Tura district.

X. The Sun Temples of Userkaf and Niuserre at Abu Gurob

1. Userkaf

Geographical position (central obelisk): 29°54'00" N, 31°11'56" E

In the proper sense, the two sun temples do not belong to our investigation program. They are not pyramids nor mortuary places, but offering temples to the solar god Re. The decision was taken to include them into the sampling process due to their close vicinity of the pyramids built by the same rulers, i.e. Userkaf and Niuserre, in Saqqara and Abusir. The question arose as to whether the building material derives from different sources, especially in the case of Userkaf, who constructed monuments at both Saqqara and Abu Gurob.

The sun temple of Userkaf was probably the first royal structure built in the Abusir area. It lies roughly half way between Abusir and Niuserre's sun temple to the north of it, near Abu Gurob, and is today in a badly destroyed condition. RICKE excavated the site systematically between 1965 and 1969 and detected various stages of construction. According to him, Userkaf only constructed the sanctuary and causeway. At a later stage, Neferirkare erected a granite obelisk and covered the base with granite and quartzite slabs. The main feature of the building is a winding corridor that leads to the roof of the base.

2. Niuserre

Geographical position (central obelisk): 29°54'15" N, 31°11'38" E

Niuserre's sun temple is located north of Userkaf's similar construction, not far from Abusir, where Niuserre's pyramid is located. It has three primary components: a valley temple, an upper temple with a courtyard, and a causeway that connected the two. BORCHARDT[124] published the building, the decorative scenes later VON BISSING/KEES[125].

According to BORCHARDT, the upper temple appears to have been built around a huge obelisk-like masonry structure.

This obelisk-like structure, built of limestone blocks, was originally approx. 36 m high. It stood on a great pedestal 20 m in height.

Near the centre of the courtyard stands a great calcite-alabaster altar formed by four "hetep" signs (i.e. a hieroglyph meaning "offering").

A number of drill holes bored into displaced granite blocks attracted the attention of several archaeologists, interested in ancient stone-working techniques. Similar holes can be found in other Old Kingdom temples mostly as door catches. Obviously, a drill-core technique was applied where rotating copper tubes shaped with a quartz-water suspension cylindrically into the rock, leaving behind a somewhat irregular hole after taking off the drill core.

Petrography

The core masonry is generally of fine-grained marly limestone, which occurs at the flanks of the Saqqara plateau. Obviously, this limestone was selectively quarried, because the coarser sandy limestone qualities are rarely incorporated into the core masonry.

On the other hand, not only the casing material and the ceiling blocks, but also many blocks of the core masonry are of the fine limestone qualities deriving from the area around Tura.

Apart from various limestone types, quartzite from Gebel el-Ahmar near to Heliopolis, granite from Aswan, and in smaller amounts greywacke from Wadi Hammamat (Eastern Desert) were used for casing the outer and decorating the inner structures. Some of the displaced granite blocks show the characteristic wedge holes, introduced as a quarrying technique during the Roman period and also employed since then, which makes a more precise dating of this kind of stone robbery impossible. Unfortunately, due to the intensive destruction of the entire site, a more precise provenance determination of the various rock types is difficult. RICKE gives valuable information on this subject[126].

In its present state, it seems that the main building material, especially that of the enclosure walls, consists mainly of mud bricks.

Macroscopically, the local building stones appear to be very similar to the Tura limestones, but they always show a distinct fine layering and a tendency to a more yellowish colour. Their geochemical attribution will be explained in connection with the Abusir pyramids (see below, page 131).

Under microscope, again the finely laminated layering, rather fossil-poor together with an always-present quartz sand detritus allows a distinction from the Tura-Maasara area, even when the fine-grained microsparitic limestone matrix greatly resembles the stone varieties from the latter area.

The provenance of the fine casing limestone, used not only for Userkaf's but also for Niuserre's sun temple, is clear according to the petrographical and geochemical results. Both analytical methods come to the same result, namely that the stone material mainly derived from the Tura region (Figs. 132–133).

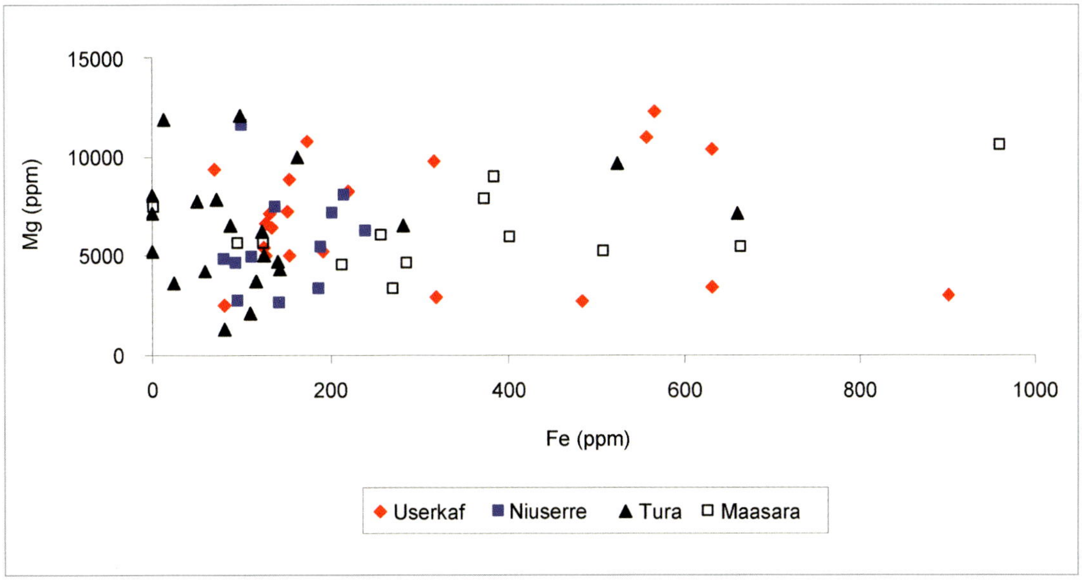

Fig. 132: Mg/Fe-diagram from the fine limestones of Userkaf and Niuserre sun temples in Abu Gurob. The plots of both temples coincide closely with the Tura quarry samples and less with those from Maasara.

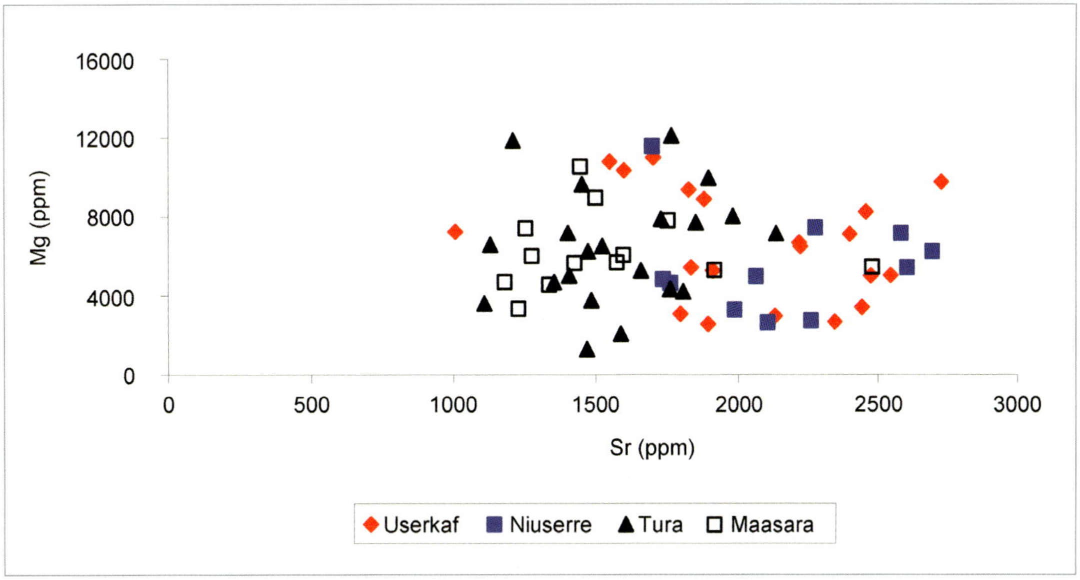

Fig. 133: Mg/Sr-diagram from the fine limestones of Userkaf and Niuserre sun temples in Abu Gurob with same result as Fig. 132.

XI. The Pyramids of Abusir

The pyramids of Abusir are situated close together on a plateau north-west of the village of Abusir, about half way between the Djoser pyramid at Saqqara and the Gizeh pyramids in the north. They all belong to kings of the 5th dynasty, the successors of Userkaf.

1. The Pyramid of Sahure (2496–2483 BC)

Geographical position (entrance of the pyramid): 29°53'53" N, 31°12'12" E

Height: 48 m | Base: 78.5 m | Slope: 50°

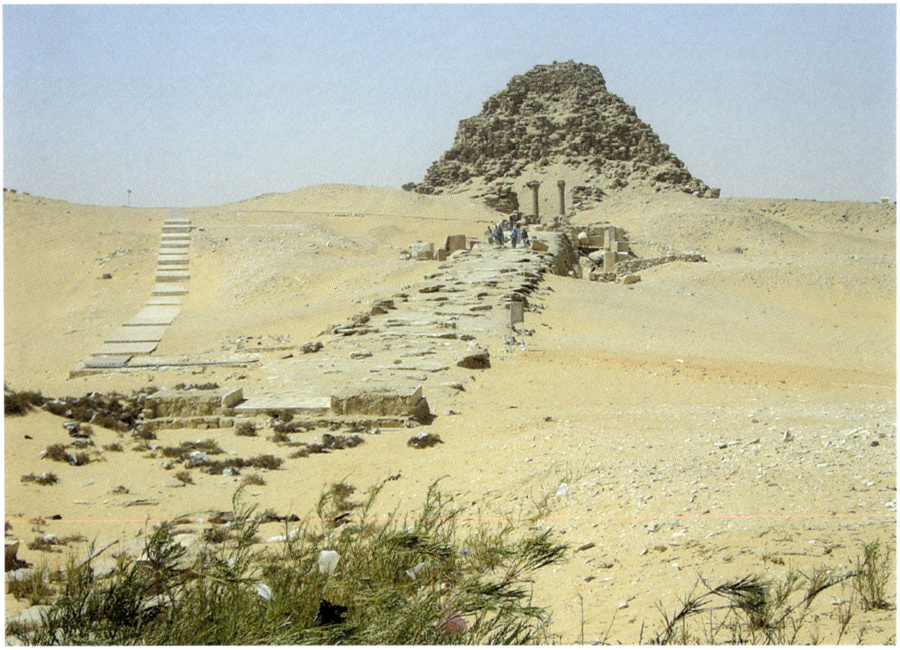

Fig. 134: Sahure pyramid with remains of valley temple and causeway (view from east).

The pyramid complex (Fig. 134) was investigated and published by BORCHARDT[127]. According to him, the pyramid was founded upon a platform made of at least two layers of limestone blocks. Originally, it had six core steps. Interspaces were filled with smaller blocks and bound together by mortar. Thus, the core is still in good condition, even though the fine limestone casing is missing today.

The entrance into the pyramid is on the north face just above ground level. A short descending corridor leads to the granite-lined passage and to the subterranean chambers. Both stand in ruins today. Impressive monolithic blocks of fine limestone form the saddle of the roof.

The valley temple has a long axis oriented north-south with a southern and an eastern ramp. Eight red granite columns adorned the temple's eastern entrance, which is paved with basalt. The central room leads to the causeway, which in turn is paved with large limestone slabs that are macroscopically identical to the core material.

The mortuary temple stands on a foundation of two layers of rough limestone blocks. The corridor is paved with basalt, and its limestone walls are decorated with relief, but only fragments remain. The outer part of the temple consists of an entrance hall and open courtyard. The courtyard once contained 16 monolithic red granite columns, standing upon a floor of irregular shaped thick basalt slabs.

A small cult pyramid stands in the south-eastern corner of the pyramid complex.

Petrography

The sandy limestones used for the core masonry of Sahure's pyramid and temple complex are predominantly of light grey and occasionally of yellowish-brown colour.

Macroscopically, the more light-coloured types feel chalky, whereas the brownish is sandier. The latter also contains visible fossil fragments that always appear grey with a glassy lustre. However, all varieties are generally very dense with distinct sedimentary bedding in parts that tend to flake off in the form of thin chips (Fig. 135).

The block sizes of the pyramid vary between 1.5–2 m in length, 0.8–1 m in width and 15–40 cm in thickness. Procuring the blocks in the quarry site, the workers followed the natural joint system, thereby avoiding unnecessary exertions needed to cut the blocks out of the solid rock. However, using this method it had to be accepted that the variation of the natural thickness of the layers and the sizes of the natural joint pattern resulted in irregular formats. Subsequently, it was more difficult to incorporate these irregular formats into an accurately planned and measured building like a pyramid (Fig. 136). It seems that the interspaces of the blocks were originally filled up by quarry waste and gravel, which has been washed off mainly at the accessible outer parts of the structure. However, parts of the pyramid mortar show a high quality, sand-rich and well-crystallized gypsum mortar.

The core material of the pyramid, the enclosure, and temple walls consist of two different types: a sand and fossil-rich hard limestone (Fig. 137) and a hard, marly limestone, which always contains sand and is very poor in microfossils (Fig. 138). At first glance, this type is very similar, disregarding the sand grains, to the Tura-Maasara limestones. When this variety gradually changes into more marly layers, it tends to flake off in a typical chip pattern (see Fig. 135).

Petrographically and geochemically, both types are identical with the limestone qualities occurring at the northern rim of the Saqqara plateau, which dips into the Abusir semi-plain. Due to the identity of the local building stones of all the Abusir pyramids, an overall geochemical description will be given below.

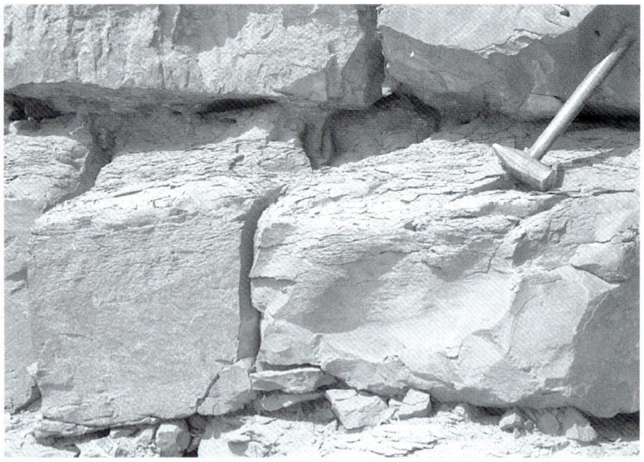

Fig. 135: Large blocks of Sahure pyramid, partly disintegrating into flakes.

▼ *Fig. 136: View from south to the Abusir pyramids with their irregular and disintegrated shapes due to their poorly formatted building material in relation to the Gizeh pyramids behind.*

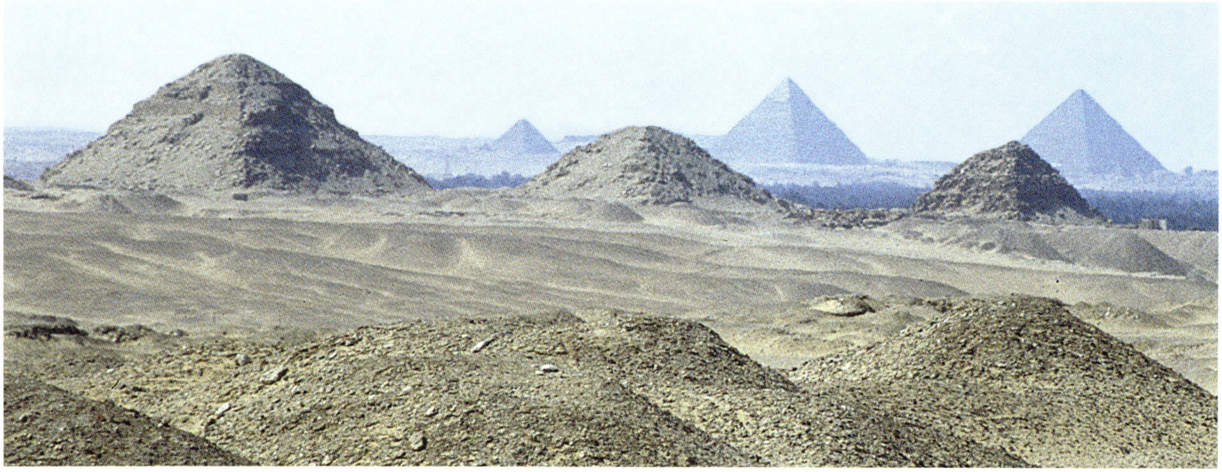

2. The Pyramid of Neferirkare (2483–2463 BC)

Geographical position: 29°53'44" N, 31°08'00" E

Height: 72 m | Base: 105 m | Slope: 53°

This pyramid was originally designed as a step pyramid with six steps (Fig. 139). The intention to finish it as a true pyramid arose during construction. This can be seen at the northern and eastern side, where the structure was built less carefully. Only a wall and a number of unsmoothed granite casing blocks exist from this second stage. Due to

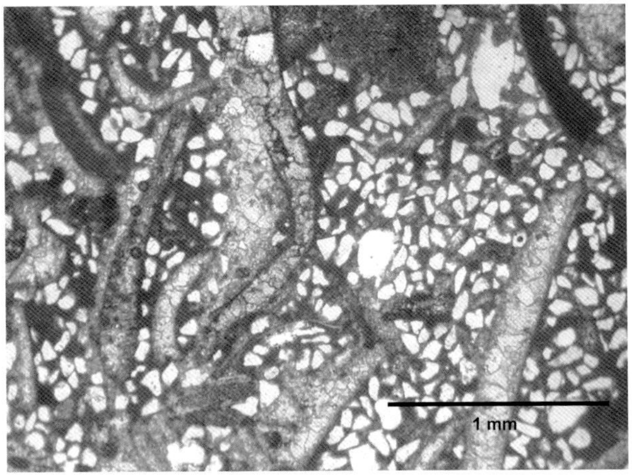

Fig. 137: Micro-photo (Sa 21). Core masonry of Sahure pyramid. Sand- and fossil-rich limestone with completely calcified fossil remains, impossible to identify exactly (1 pol. filter).

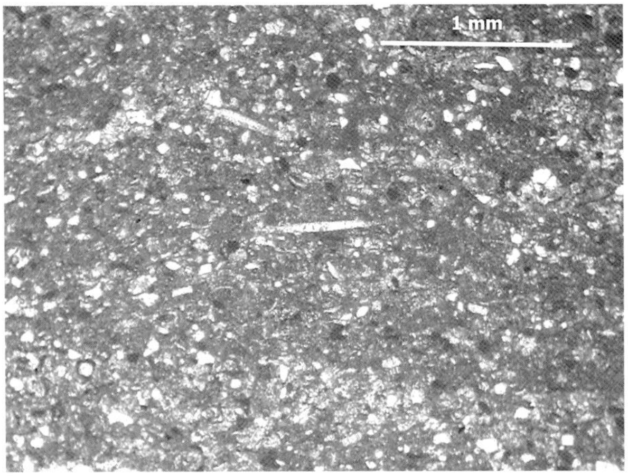

Fig. 138: Micro-photo (Sa 1). Core masonry from Sahure pyramid. Fine-grained, marly limestone with many fine sand grains (white) and only few micro-fossils like sponge needles (1 pol. filter).

the presence of wedge holes, these granite blocks were partly reworked during the Roman period or later. However, both the pyramid and its temples were most probably never completed.

Neferirkare's pyramid is the largest and best preserved of the Abusir group. It was built with different sized blocks of local limestone comparable to Sahure's pyramid. The six steps and the core itself were cased with large blocks of fine limestone, as was probably the outer surface of the monument with the lowest course of red granite blocks.

The valley temple, now in total ruin, shows remains of twelve columns. Blocks of granite, basalt, and fine limestone give evidence of its former splendor.

The causeway is paved with basalt blocks; the walls above it are of fine limestone. Later, king Niuserre diverted the upper end of Neferirkare's causeway to his own pyramid.

The mortuary temple was built predominantly using mud bricks and had a floor of packed clay and fine limestone slabs without a solid foundation. Granite columns once adorned the courtyard, but only little of them remain, as this material was removed and presumably reused. On granite pieces, traces of wedge holes are clearly observable.

Originally, a great mud brick wall surrounded the whole complex, though parts of it are barely distinguishable today.

The whole complex was investigated and published by BORCHARDT[128].

Petrography

The walls of the pyramid's steps were made of large blocks with sizes up to 2 m long, 0.5–0.8 m high and around 1 m deep alternating with blocks 0.7–0.9 m long, 0.15–0.2 m high and 0.4–0.6 cm deep and of local sandy limestone (see Fig. 139). Towards the inner parts, the masonry was constructed more negligently and with smaller block sizes.

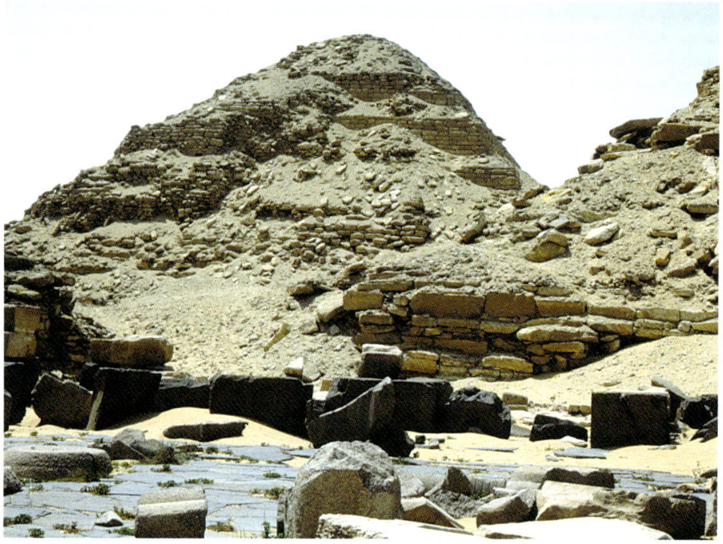

Fig. 139: Neferirkare pyramid in Abusir (left) seen from north-east. Despite heavy decay, the steps are still well recognizable. In the foreground are the remains of Niuserre's mortuary temple with its basalt pavement.

The local building stones used in the core masonry are again of the bimodal character similar to the stone composition throughout the entire Saqqara plateau and at its northern edge, which extends into the Abusir plain. Especially the larger blocks consist of a sandy limestone, which is mainly rich in calcified fossil fragments and partly also with bituminous matter. The second quality is a marly limestone with less and smaller quartz grains, and relatively poor in fossil fragments, mostly indiscernibly milled. This type seems to be less weather-resistant.

At the northern entrance of the pyramid, both the local building stones as well as a fine limestone are used in large blocks. The same fine limestone is found in some of the surrounding temple walls. In both cases, it is the typical grey-white, dense limestone from Tura-Maasara.

3. The Pyramid of Khentkaus II

Geographical position: 29°53'39.5" N, 31°12'08" E

Height: 17 m | Base: 25 m | Slope: 52°

A small pyramid structure lies on the south side of the pyramid of Neferirkare (Fig. 140). Its true nature was recently revealed and its owner could be established as Khentkaus II, the wife of Neferirkare and the mother of "two kings of Upper and Lower Egypt" namely Neferefre and Niuserre[129].

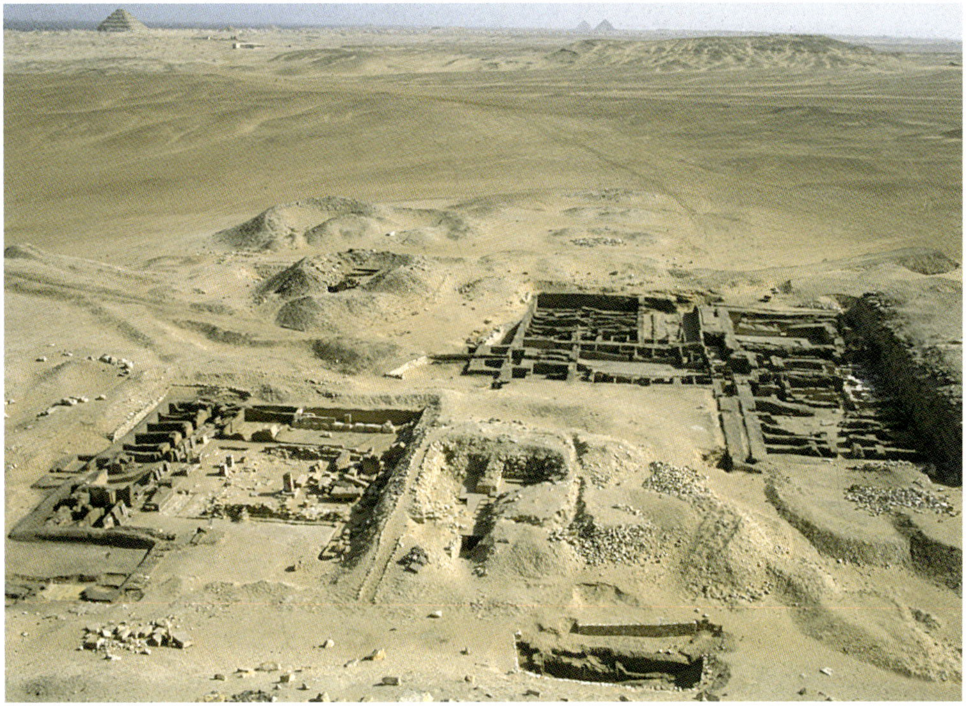

Fig. 140: Overview of the small Khentkaus II pyramid with its adjacent temple complex on its east side, Neferefre's temple behind. View from top of Neferirkare pyramid.

The pyramid was probably begun during the reign of Neferirkare, but due to the king's death, the construction was halted. After some time, the pyramid was finished by Khentkaus' son, Niuserre.

The pyramid itself, which lies mostly in ruins, was long overseen by archaeologists. It was built using the discarded limestone from the nearby Neferirkare pyramid, which is consequently of the same building material[130]. The core shows three steps, bound with clay mortar. The casing was of fine limestone, and it had once been capped by a dark granodiorite pyramidion, of which fragments have been found.

The entrance to the pyramid is near ground level in the middle of the north wall. In the pyramid, remains of the queen's burial equipment were found, including fragments of a red granite sarcophagus.

The mortuary temple that stands in front of the pyramid's east wall was finished in several stages: the earliest part of the temple is made from limestone and was entered through a pillared portico from the south-east. The second phase of the construction was mostly of mud brick. Further, a small cult pyramid was constructed in the south-east corner of the mortuary temple.

4. The Pyramid of Neferefre (2456–2445 BC)

Geographical position: 29°35'39" N, 31°12'05" E

Height of unfinished structure: 7 m | Base: 65.5 m | Slope: 64°

This most south-western pyramid remained unfinished (Fig. 141), and apparently, only the lower most core of one-step was finished before the king's early death. Before excavation, the pyramid was thus regarded, due to a mud brick cover, as a low mastaba. Only after thorough examinations could it be identified as Neferefre's pyramid[131].

The core was built of five horizontal layers, about 1 m high, with the outer mantle consisting of large, rough blocks up to 5 m long and an inner core of less carefully laid, smaller limestone blocks. These were stacked to make the first core step about 7 m high. A clay mortar was used for binding. An interspace between the inner and outer layers of

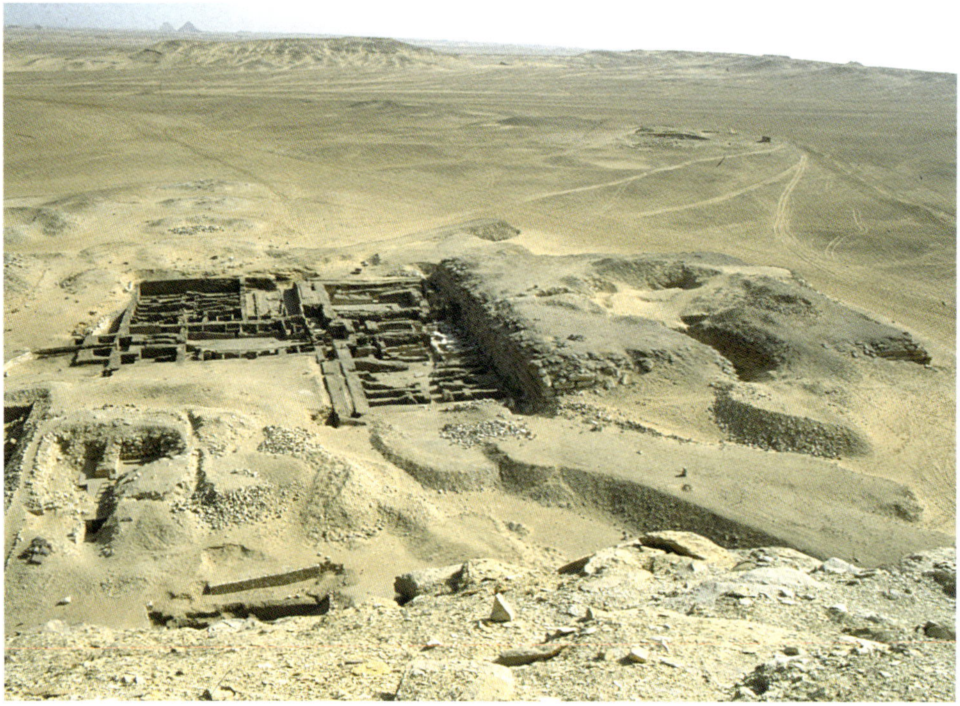

Fig. 141: Remains of the Neferefre pyramid with its adjacent temple complex on its east side. View from top of Neferirkare pyramid.

the core was filled with sand, rubble, clay, and stone fragments. Neferefre probably died prior to the completion of the first core level, and so the pyramid as well as the mortuary temple were presumably finished by his successors.

The entrance to this pyramid is in the middle of the north side, close to ground level. It curves slightly to the south-east before reaching the antechamber, which was constructed on a lower level and lined with red granite.

On the eastern side of the same foundation on which the pyramid is constructed, a small mortuary temple, aligned north-south, was built of smaller, fine limestone blocks. The entrance was in the south-east. During the reign of Niuserre, additions were made to the temple along the east side of the pyramid in the form of a large mud brick section.

The southern part of this additional structure consists of an east-west oriented hall with 20 wooden lotus columns. This was the first hypostyle hall that we know from the age of the pyramid builders.

The whole of the temple was enclosed within a huge mud brick wall that was reinforced with limestone monoliths at the corners.

Niuserre also extended the temple complex on the east side. In this case, the additional architectural elements included a monumental entrance with two limestone papyrus columns, an entrance hall, and an open courtyard with 22 wooden columns.

Since most of the mortuary temple was heavily covered by waste and sand, a collection of mostly wooden statues and a papyrus-archive has remained well-preserved[132].

The building stones of the Neferefre Pyramid

Again, the building stones are of the two types of limestone already described above. The casing, as far as only some remains indicate, was of fine limestone from the Tura-Maasara area. A more precise localization is not permissible due to the limited samples available.

However, the floor of large parts of the mortuary temple was paved with fine limestone, indeed, the same type as could be found at the pyramid's wall next to the mortuary temple. Somewhat enigmatic are several well-polished blocks used in the pavement that consist of limestone with fossils of *Nummulites gisehensis*. The nearest occurrence of the latter is at the Gizeh plateau and was procured most probably from the Menkaure quarry site. This might be an indication of reuse of the Gizeh quarries during the 5th dynasty.

5. *The Pyramid of Niuserre (2445–2414 BC)*

Geographical position: 29°53'46" N, 31°12'12" E

Height: 52 m | Base: 79 m | Slope: 51°

It seems as if Niuserre wanted to erect his pyramid (Fig. 142) in the close vicinity of his family members. This would explain the unusual position next to the mortuary temple of his father Neferirkare and directly in front of the mastabas belonging to members of the royal court.

It was again BORCHARDT[133] who first carried out a systematic investigation of the site.

The pyramid has a core of seven steps built with local limestone. It was then completed as a true pyramid, the outer mantle of which was made of fine limestone.

The entrance was in the middle of the north side at ground level. The entrance opens to a corridor lined with fine limestone and reinforced with red granite at each end. A barrier consisting of two granite blocks divided this corridor.

The mortuary temple on the east side of the pyramid is rather small, and unlike the other pyramids at Abusir, it is shaped in the form of an L.

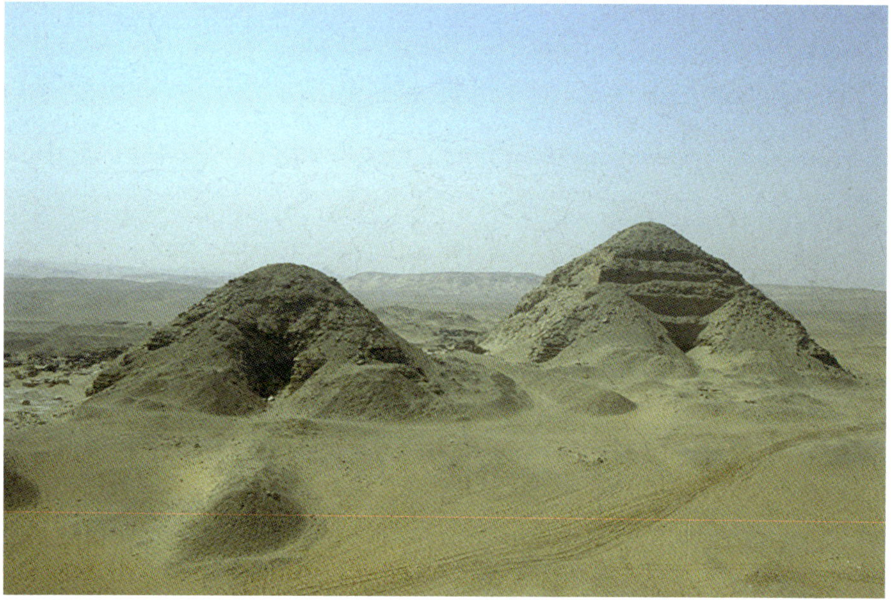

Fig. 142: Niuserre (left) and Neferirkare pyramid in Abusir, seen from top of Sahure pyramid.

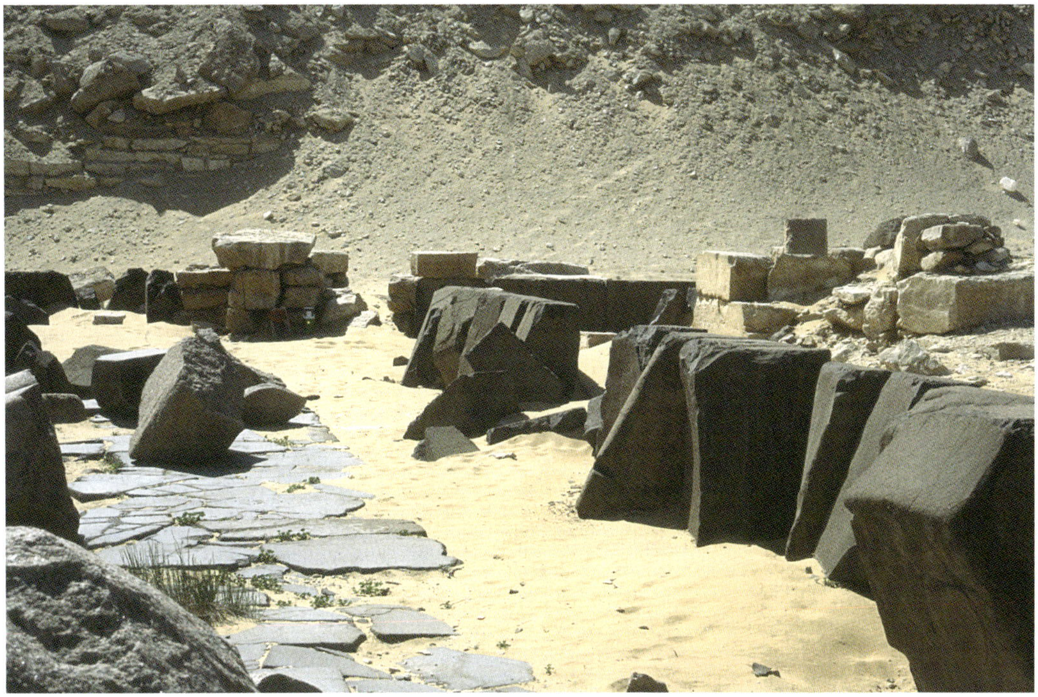

Fig. 143: Basalt pavement and block walls (right) of the Niuserre mortuary temple. Due to a long time coverage with sand and temple debris, most of the basalt blocks still remain in a rather good state.

The entrance corridor leads to a courtyard that was also paved with basalt and where basalt blocks of up to 1.5 m in height were erected to form the walls (Fig. 143). The 16 papyrus columns along the walls as well as the door gates were made of red granite in various tints. Unfortunately, most of them were destroyed by wedge splitting in later periods.

The discovery of calcite-alabaster fragments indicates that an altar once stood in the north-west corner of the courtyard.

The pyramid stands in a limestone-paved courtyard. Again, a cult pyramid lies in the south-eastern corner of the precinct.

The building stones of the Niuserre pyramid

The buildings stones of the core material are exclusively of local provenance. Again, they consist of the two types described above: massive sand and fossil-rich limestone and a fine-grained marly limestone with less and finer-grained sand components. The last type was clearly preferred for the masonry, which explains the more advanced stage of

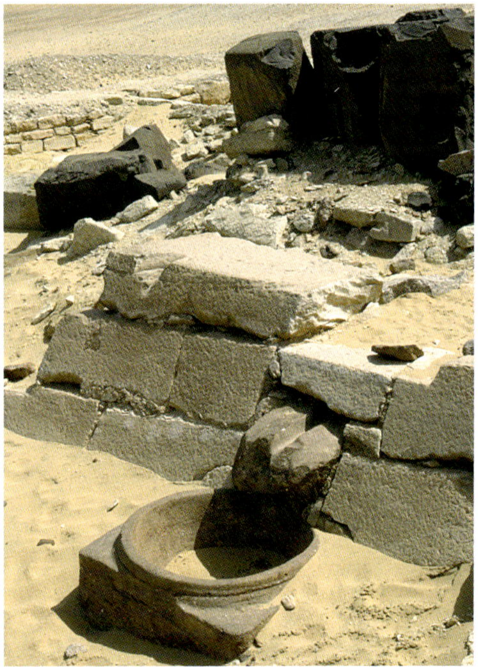

Fig. 144: Niuserre's pyramid temple: Remains of the drainage system (rim and basin) made of quartzite.

deterioration of the entire construction. The block sizes of the steps' outer walls are up to 1.5 m long, between 0.2 and 0.8 m thick, and 1–0.8 m deep.

The temple walls are also of this local material, and the blocks are large in size, i.e. up to 2.0 × 1.5 m. Fine limestone from Maasara was also used in the temple architecture. In the latter parts, the decoration is especially well-preserved.

The basalt blocks, when exposed to the sun, tend to become brittle due to the thermal fluctuations, heating up during the day and cooling down at night. This process is less effective, when basalt blocks are covered in a building and/or when they have direct contact with the soil, e.g. as paving blocks. Due to high thermal conductivity, basalt transfers the sun-acquired heat relatively fast to the soil (see Fig. 143).

Other interesting architectural details are the drainage installations, which are exclusively made of quartzite, mainly a conglomerate variety (Fig. 144). Further, many fragments of normal quartzite were recovered in the excavation debris.

Geochemical provenance determination of the core masonry of the Abusir pyramids and the sun temples of Abu Gurob

In the immediate vicinity of the pyramid complex at Abusir, neither quarry sites nor any significant outcrops could be detected. The pyramid constructors were dependent on distant sources for their entire building material, as the closest outcrops of suitable stone quality occur only at the northern escarpment of the Saqqara and Abusir plateau

Fig. 145: Part of the quarry outcrops at the northern flank of the Saqqara plateau towards the semi-plain of Abusir.

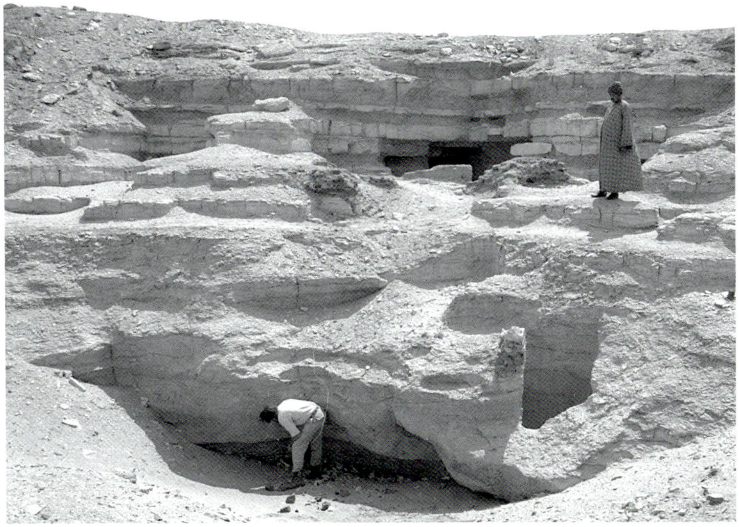

Fig. 146: Mg/Fe-diagram of core masonry from all Abusir pyramids. With only a few exceptions, mainly of Niuserre, they all plot within a distinct geochemical distribution field, which coincides closely with the respective field of the Abusir and Saqqara-North quarries.

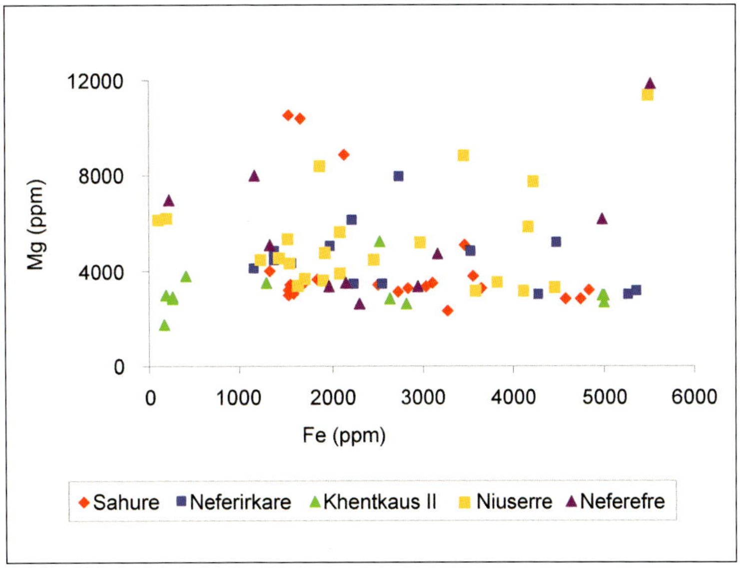

(Fig. 145). Generally, the same geological units occur here along the eastern escarpment as already discussed. In contrast to the eastern escarpment, only slight lateral variations, but no significant geochemical differentiation can be expected. Indeed, the geochemical correlation diagrams show a striking similarity between the Abusir pyramids. Especially the Mg/Fe-diagram (Fig. 146) and that of the Abusir-Saqqara-North quarry sites (Fig. 147a) indicate a close geochemical relation and the Mg/Sr-diagrams (not given here) generally show the same tendency.

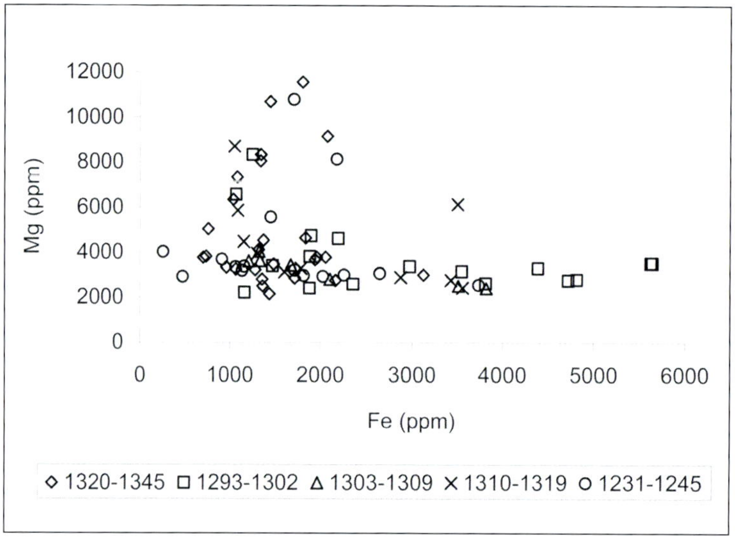

Fig. 147a: Mg/Fe-diagram of the Abusir and Saqqara-North quarry sites. Note the close coincidence with the Abusir pyramids core material in Fig. 146.

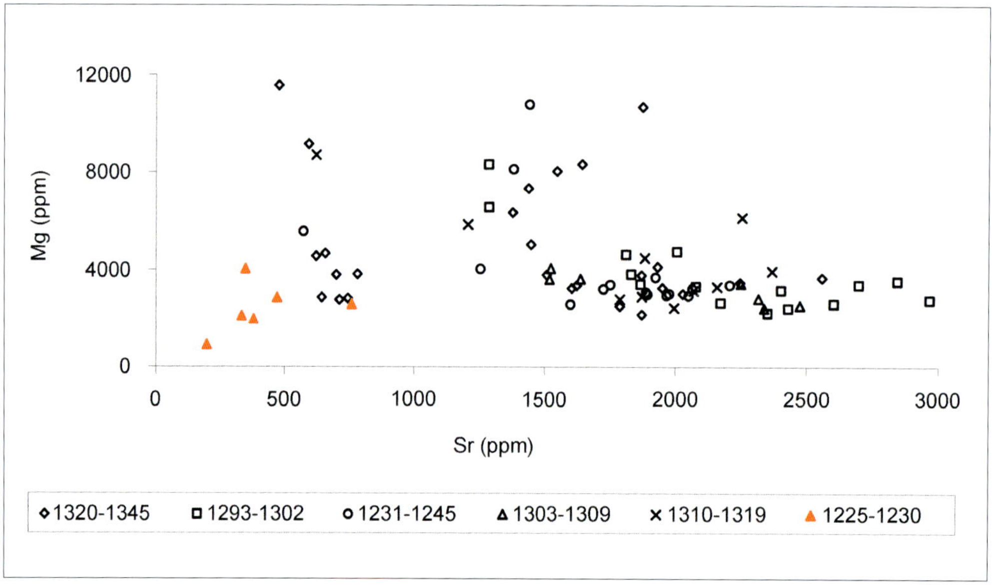

Fig. 147b: Mg/Sr-diagram of the Abusir and Saqqara-North quarry sites (small plots) and the quarry area assumed by VERNER *(2005) as a source of the Abusir pyramids (black triangles). The totally different petrographical appearance and the separated geochemical position of the "Verner samples" to the Abusir and Saqqara-North quarries hardly support this assumption.*

Recently, VERNER suspected an area further west of the road between Abusir and the Saqqara rest house to be the source of the Abusir core building stones and listed it on his map[134]. However, he seemed uncertain and asked for detailed scientific investigations. Thus, this area was encompassed into the present study and six specimens were sampled. Despite the small number of samples, the petrographical appearance was completely different in relation to the pyramid core samples. Geochemically, they also differ significantly (Fig. 147b). However, the area could not be intensively investigated for other potential quarry sites. Yet, VERNER describes stone implements, traces of wedge splitting on some fragments, and even chisel marks in this area. According to his short remarks, a more recent quarry site seems to be more plausible. Nevertheless, this should be confirmed by further investigation.

Finally, the data of the sun temple's core masonry should be discussed. A sufficient number of core masonry samples could only be taken from the better preserved sanctuary complex of Niuserre. According to the geochemical correlation diagrams, these give a clear provenance attribution to the Abusir quarry sites of the northern escarpment. A precise attribution for the core masonry of the Userkaf sanctuary complex is difficult. In this case, the very poor sample data do not allow an unambiguous statement and are consequently not listed in Fig. 146. Only indirectly may the conclusion be drawn that this material derived from the Saqqara-North quarries along the eastern escarpment as the geochemical data from monuments and quarry sites dating to Sahure's reign, when the northern Abusir quarry sites had not yet been opened.

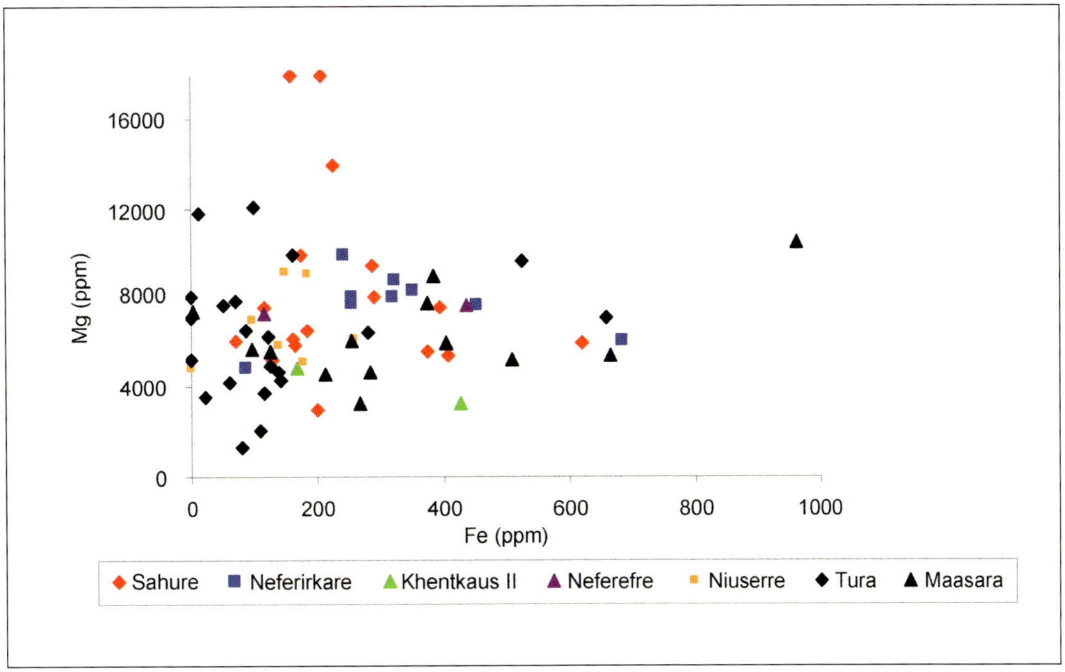

Fig. 148: Mg/Fe-diagram of casing material from the Abusir pyramids. In spite of a few deviations, especially of Sahure, a sufficient good coincidence with the Tura-Maasara province is obvious.

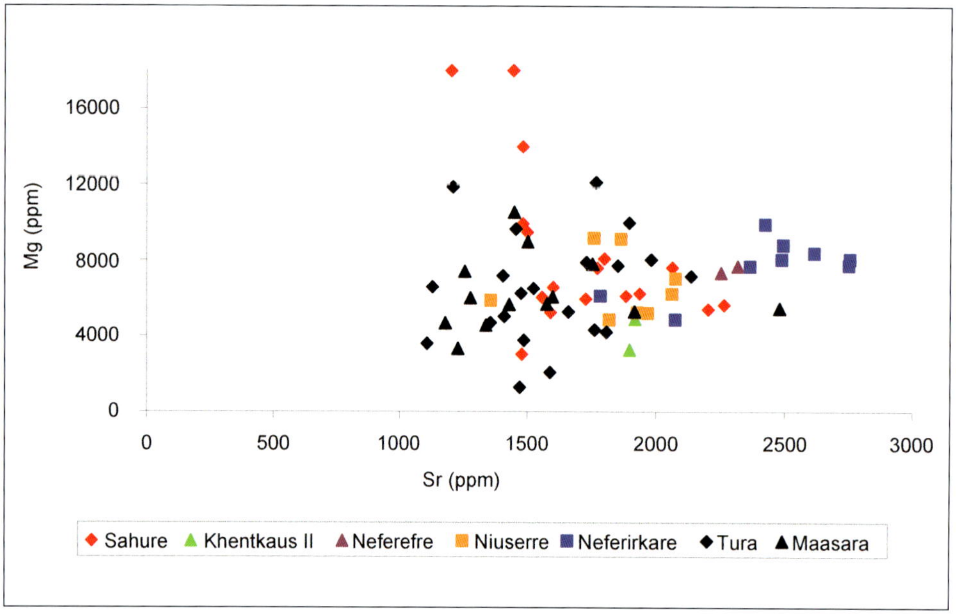

Fig. 149: Mg/Sr-diagram of casing material from the Abusir pyramids. A sufficient good correlation with the Tura-Maasara (gallery) quarry province, even with a certain tendency to slightly elevated Sr-values, is striking. Only for Niuserre a preference to Tura is evident.

Geochemical provenance determination of the casing material of the Abusir pyramids and the sun temples of Abu Gurob

According to the geochemical correlation diagrams of Figs. 148–149, the casing limestones of all the Abusir pyramids and the sun temples derive from the Tura-Maasara quarry areas. However, the distribution of the sample values from the various areas differs within the Tura-Maasara field so significantly that an individual gallery quarry may be expected for each royal complex.

The data from the casing limestones of Neferirkare and Neferefre's monuments plotted within almost identical fields, so that it can be assumed that the two rulers worked gallery quarries situated close to one another, adhering to the rule that each king opened his own quarry.

XII. The Pyramid of Djedkare-Isesi (2405–2367 BC) at Saqqara-South

Geographical position (entrance): 29°51'05" N, 31°13'14" E

Height: 52.5 m | Base: 78.5 m | Slope: 52° | Length of causeway: 220 m

The pyramid originally had a core of six steps, but the upper three steps no longer exist. Today, the pyramid is only about 24 m high (Fig. 150). Each step is approx. 7 m high and built with small, irregular blocks of local limestone bound with clay mortar. Most of the casing is missing. The entrance to the pyramid is on the north side, located in the pavement of the courtyard. The burial chamber once had a sarcophagus of greywacke and calcite-alabaster canopic jars, of which only fragments were found.

The ruins of the valley temple, if existing, lie under the modern village of Saqqara. Remains of blocks with relief decoration and a few red granite pieces are scattered around the houses there.

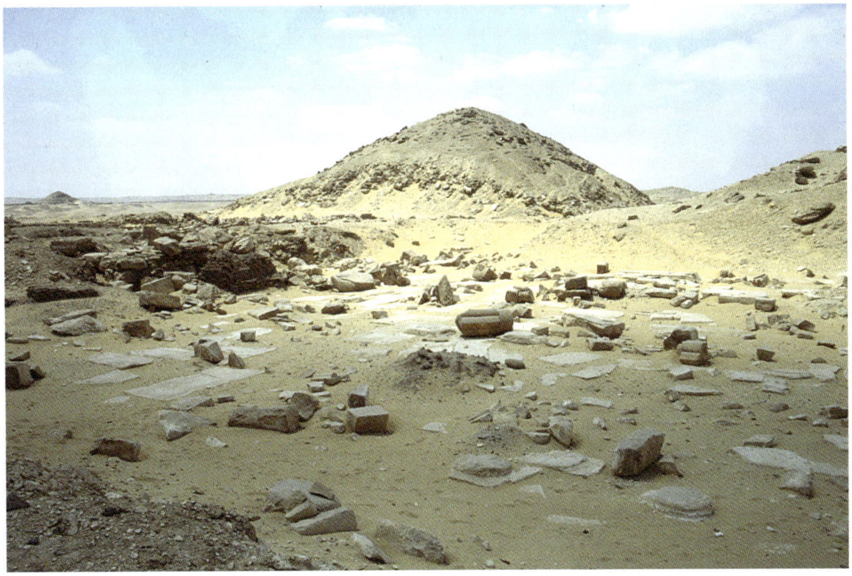

Fig. 150: Djedkare Isesi's pyramid, view from north-east, with queen's pyramid in the foreground (left).

The causeway deviates slightly to the south from the east-west axis of the mortuary temple, which is situated to the east of the pyramid. The mortuary temple is badly damaged today. At the entrance of the temple stood two massive, tower-like structures, which may be the forerunners of the pylons of later temples. The entrance hall was probably paved with calcite-alabaster, which continued into the temple courtyard.

The small cult pyramid near the south-east corner of the main pyramid is very similar to others built during this period. It was constructed with three steps and a single underground chamber. The cult pyramid was enclosed within a small perimeter wall at the south-east edge of the pyramid.

Further, the pyramid of an unknown queen adjoins the north-eastern corner of the temple wall (see Fig. 150).

An Egyptian team[135] carried out the most recent investigations at the site.

Petrography of the core material and the assumed quarry sites of the Saqqara-South pyramids

All of the various pyramid complexes and the assumed quarry locations at Saqqara-South were accessible for sampling. During the geochemical and petrographical investigation, it became clear that the core masonry of most of the pyramids generally derives from Saqqara-South sites near to the monuments, but a more precise attribution to well-defined quarries was not possible. Therefore, a reinvestigation of this area was arranged. However, during this second campaign the Egyptian Antiquity Organization (EAO) first withdrew the export licence for scientific samples in general and finally prohibited sampling totally. Thus, the provenance attribution is only based on this previous and somewhat incomplete sampling.

The most probable quarry sites during the first sampling campaign in 1986 were the northern flank of Wadi Tafla and the eastern escarpment of the Saqqara-South semi-plateau towards the Nile valley. The latter escarpment seems to have been artificially straightened and is only interrupted by Djedkare-Isesi's causeway. Interestingly enough, the samples procured from these sites are petrographically identical to those from all the Saqqara-South pyramid core stones. However, it has to be considered that the outcropping rock formations from the Abusir plain and northern flank of Saqqara, which extends to Dahshur, belong to the same geological unit, described in detail by YOUSSEF ET AL.[136].

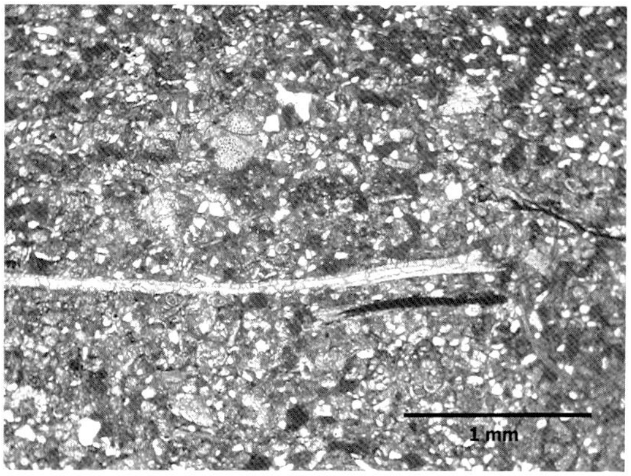

Fig. 151: Micro-photo (Dj 26) of core limestone from the Djedkare Isesi pyramid. The sandy limestone contains many small fossil remains, mainly calcified, and calcified algae filaments, partly filled with bituminous mass (black) (1 pol.filter).

Consequently, in Saqqara-South these rock units also consist generally of two types of sandy limestone. One type, mostly yellowish-grey, has distinct sedimentary bedding, feels like very fine-grained sandstone, and macroscopically always contains greyish, glassy fossil fragments that are impossible to identify. The second type is grey-white in colour, has less-evolved bedding, and has a similar chalky feel as a marly limestone but with no recognizable fossils.

Under microscope, the first type is rich in quartz sand grains, 0.2–0.01 mm in size, set in a sparitic matrix containing many fine-milled fossil relicts and occasionally fish scutes. Larger fossil remains, mainly up to 5 mm large calcified shells and calcified algae filaments, often with a characteristic mass of black bitumen at their rims, dominate with small gastropods the fossil components (Fig. 151). The more marly second type is of a microsparitic matrix with many fine-milled fossil fragments, but poor in sharp-edged sand grains with sizes ranging between 0.005–0.05 mm. Under higher magnification remains of globigerinae, coccoliths, and sponge needles are detectable in the matrix. In both types, dark appearing oval peloids are sporadically present.

Geochemical determination of the core material from the Saqqara-South pyramids

Comparing Figs. 152 and 153, which show the elements Mg/Fe and Mg/Sr, most of the samples taken from the core masonry of the Djedkare-Isesi, Merenre and Pepi I pyramids correspond well to the "quarry" samples along the eastern escarpment and at the northern flank of Wadi Tafla. However, the geochemical data from the core masonry

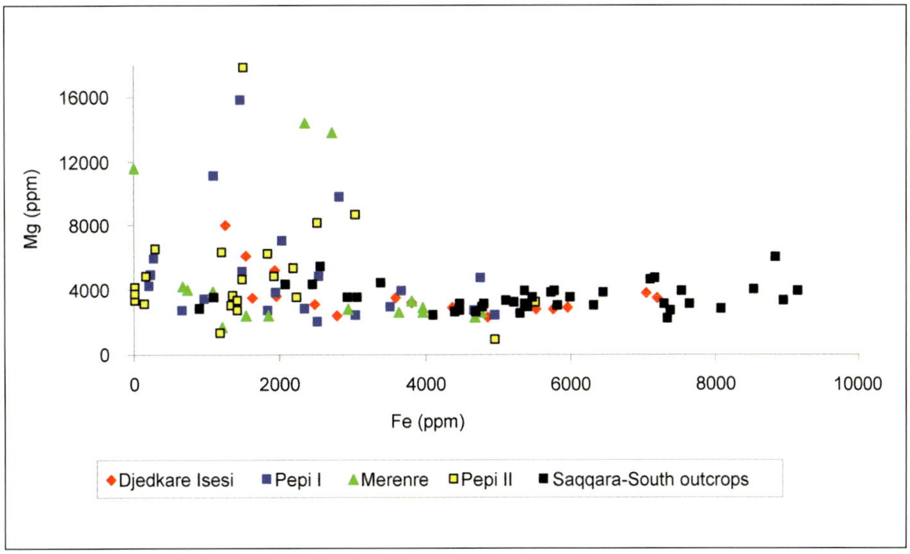

Fig. 152: Mg/Fe-diagram of Saqqara-South pyramids core masonry and the nearby outcrops. Only for the pyramids of Djedkare Isesi, Merenre, and Pepi I most of the samples coincide rather satisfactorily with the outcropping sandy limestones of the environment, but only poorly with those from Pepi II.

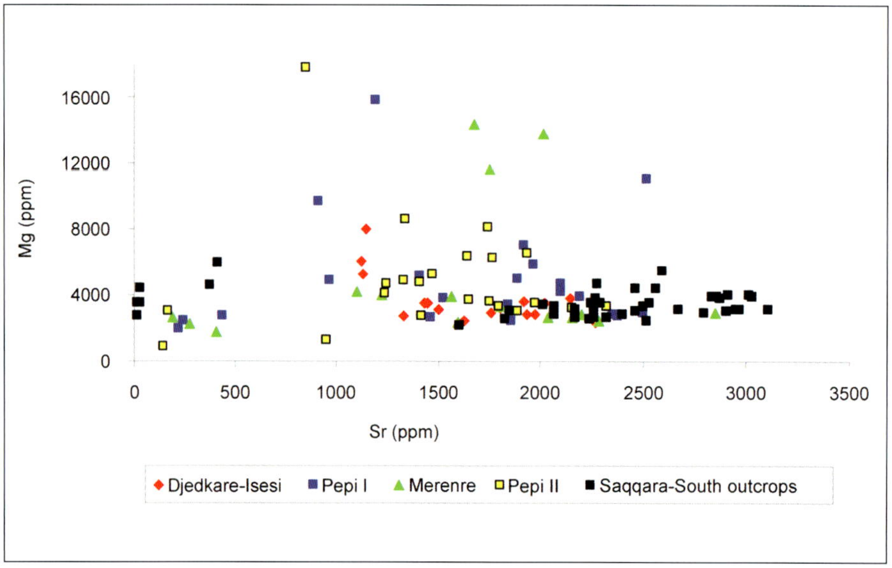

Fig. 153: Mg/Sr-diagram of the Saqqara-South pyramids' core masonry and the nearby outcrops. The overlapping between pyramid masonry samples and the local outcrops is about the same as in Fig. 152.

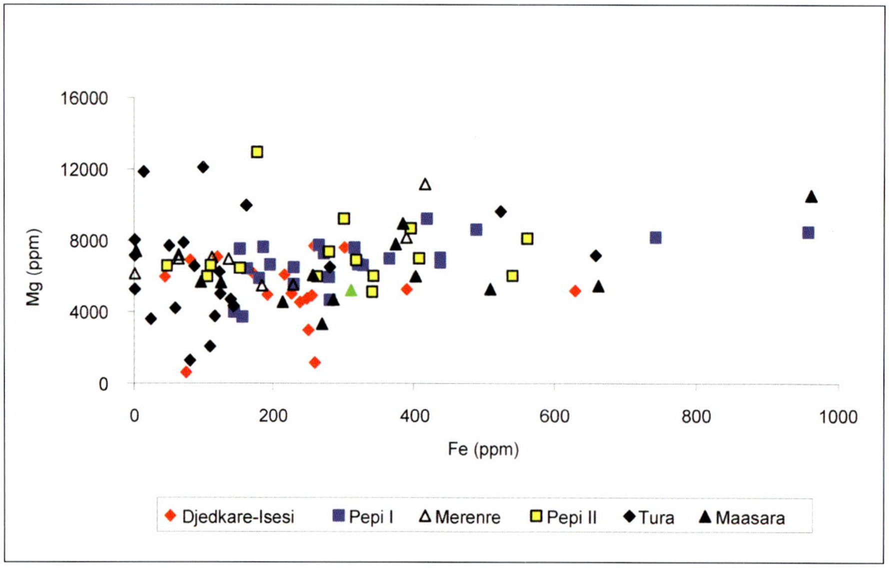

Fig. 154: Diagram Mg/Fe of the Saqqara-South pyramids' casing masonry in comparison with limestones from Tura and Maasara. A convincing differentiation between the two localities is not possible.

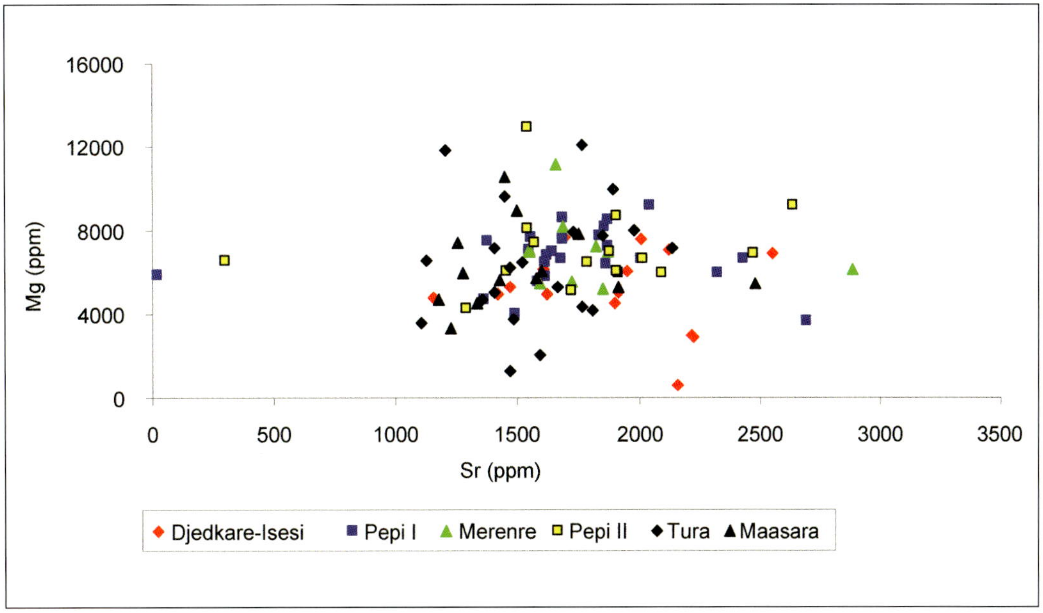

Fig. 155: Mg/Sr-diagram of the Saqqara-South pyramids' core masonry in comparison with limestones from Tura and Maasara. No distinct differentiation between Tura and Maasara is possible.

of Pepi II is only vaguely similar to these quarry sites. On the other hand, the petrographical structure of the rocks and their sparse fossil content conforms to the core material of the Djedkare-Isesi, Merenre and Pepi I pyramids. Again, it can therefore be assumed that the exact quarry site is situated within close range of the pyramids. Despite intensive searching this assumption could not be confirmed during the sampling campaign.

Geochemical determination of the casing material of the Saqqara-South pyramids and temple complexes

Figs. 154–155 demonstrate that the casing limestone of both the sparsely remaining original pyramid casing and the various temple areas derived from the Tura-Maasara quarry district. A more detailed comparison of the Mn/Fe- and Mg/Sr-diagrams supports a more exact attribution to the quarry sub-district of Maasara. This premise also fits in regards to the entire Saqqara-South pyramid field. This, of course, is not surprising, as this sub-district lies almost directly opposite Saqqara-South.

XIII. The Pyramid of Unas (2367–2347 BC) at Saqqara-North

Geographical position (pyramid entrance): 29°52'07" N, 31°12'53" E

Height: 43 m | Base: 57.75 m | Slope: 56°

The core of the pyramid consists of six steps with rough blocks of local limestone that gradually decrease in size towards the top layer (Fig. 156). The casing was of fine limestone, some of which remains *in situ* on the lower levels.

The design of both the substructure as well as Unas' mortuary temple is very similar to his father's (Djedkare) complex, whereby the original entrance was located under the north chapel. The north chapel no longer remains.

Inside, there are corridors leading to an antechamber and a burial chamber, which is lined with calcite-alabaster. In both of the chambers, the "pyramid texts" were carved in bas-relief and the individual hieroglyphs were painted blue-green on all but the west wall of the burial chamber.

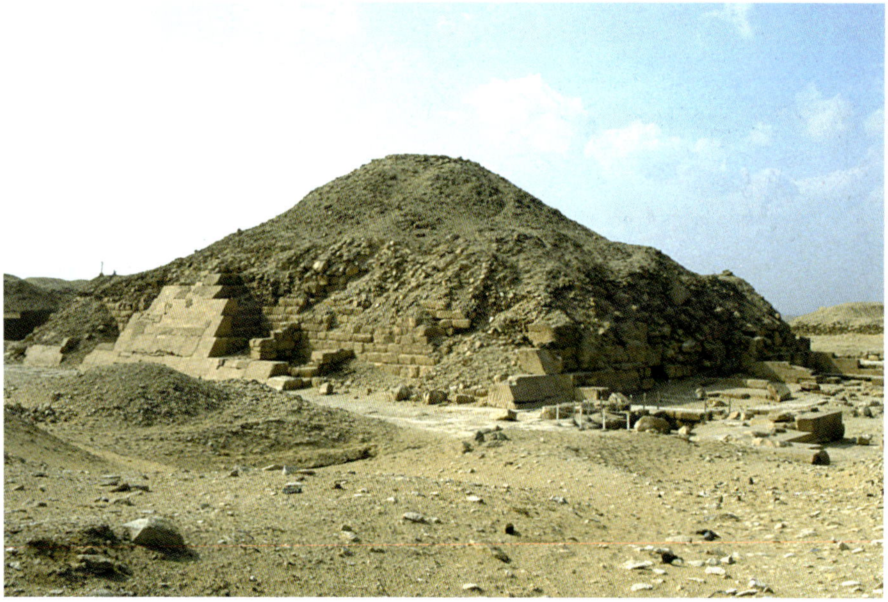

Fig. 156: Unas pyramid from south-east. In spite of the ruined state, remains of the original casing still exist at the base of the pyramid.

In the frame of the present investigation, it clearly could be ascertained that the sarcophagus is made of greywacke and not of basalt, as often reported.

The long causeway shows two bends, probably intentionally constructed to avoid other buildings. In fact, material from older buildings was used in the pavement. Bas-relief scenes adorn the walls of the causeway. They show, amongst other things, boats transporting granite palm columns from Aswan intended for Unas' monuments. Two of them are still standing in his valley temple.

The causeway leads through a red granite gateway to the calcite-alabaster paved entrance hall of the mortuary temple situated east of the pyramid. 18 red granite columns surrounded an open courtyard.

Aside from a red granite false door, little else remains of the mortuary temple today.

The whole pyramid, including a small cult pyramid in the south-eastern corner, was once surrounded by a massive stone enclosure wall.

Petrography

The inner core of the Unas pyramid was built with relatively small sizes of stone between 40 to 60 cm in length, around 15 cm in thickness and 20–40 cm in depth. In addition to this small format, block sizes of up to 2.5 × 0.5 × 0.6 m were used in the area of the casing stones at the pyramid's lower layers, thus giving the outer parts more stability (Fig. 157). The core predominantly consists of stone material extracted from sources along the escarpment towards the Nile valley. This observable fact has already been described for Djoser's, Unas' and the other pyramids of the Saqqara plateau.

However, during core sampling, an additional stone type stood out even on a macroscopical level. This variety resembles the fine stone quality used for the casing of the Djoser and Sekhemkhet pyramid complexes. This correlation was then proven petrographically. It could indeed be confirmed that the fine limestone qualities, occasionally incorporated into the core masonry of the Unas pyramid, are of the very same type as the casing stones used for the Djoser and Sekhemkhet pyramid complexes.

The pyramid's casing is only partly preserved. According to an inscription at its southern side, intensive restoration work was undertaken at this royal monument as early as the New Kingdom, under the supervision of Khaemwese, a son of Ramses II.

The Saqqara core material has already been described above and does not need repeating here.

Both the casing stones of the Unas pyramid as well as the wall blocks within the subterranean chambers and those of the entrance corridor are of a fine and weather-resistant limestone similar to the stone, which occurs at the Tura-Maasara quarry sites.

Under microscope, both the very fine limestone, which mainly derives from Tura, and the coarser-grained type, which is predominantly from Maasara, could be identified. However, the geochemical data rather indicate Maasara. Therefore, it is very probable that this very fine limestone also occurs there.

The mortuary temple and the roofed causeway were mainly constructed with fine limestone from, most probably, Maasara. However, the inner parts of the temple had an alabaster floor and the columns in the central court are of both quartzite from Gebel el-Ahmar, near Heliopolis, and of red granite from Aswan. Other architectural elements such as thresholds and architraves were constructed using three varieties of granite (pink, dark red and grey granodiorite) that all occur south of Aswan.

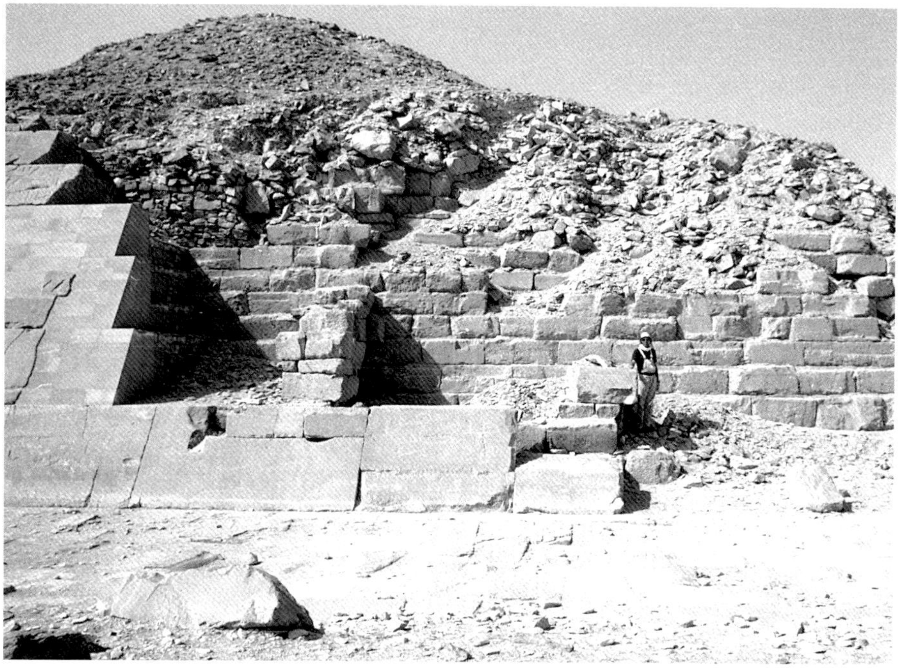

Fig. 157: Unas pyramid, southern side. Note the large casing blocks and the well formatted blocks of local material behind them relative to the irregular small sizes of the inner core.

Geochemical provenance determination of the core material

By far the largest part of the core limestones can be geochemically attributed to the quarry area located at the escarpment sloping down to the Nile valley, which today is mostly buried by debris (Figs. 158–159). However, quite a number of samples plotted outside the field of the Saqqara escarpment quarries, which could not be explained by statistical exceptions only.

It is striking that some of these samples are very poor in Fe-content (Fig. 158). This is characteristic of limestone types previously used by Djoser and Sekhemkhet for casing. It should be recalled that this weather-resistant material was also used at Meidum and in the small pyramid of Huni (?) at Zawiyet el-Meitin (Zawiyet el-Sultan). As outlined above, the limited use of this limestone during older periods was presumably due to exhaustion of the source material, which resulted in the opening of the Tura-Maasara quarries under Snofru. After almost 250 years, this building material appears again, this time, however, confined to the core masonry. It is reasonable that this material constitutes reused stones from ancestors' monuments. The then still unburied northern part of Sekhemkhet's pyramid complex would be the most likely and nearest source. This possibility is all the more plausible, as the unfinished state of this almost forgotten pyramid would have invited subsequent rulers to reuse the stone material without any significant qualms of conscience.

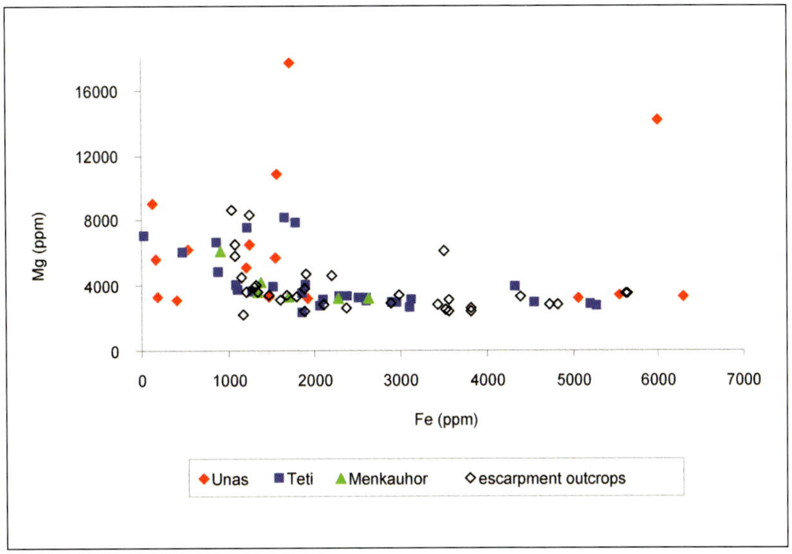

Fig. 158: Mg/Fe-diagram of the pyramids of Unas, Teti, and Menkauhor in comparison with the assumed Saqqara quarry sites. A generally good coincidence of the plots is evident with the exception of three Unas samples below 500 ppm Fe, which are most probably spolia from Sekhemkhet's district.

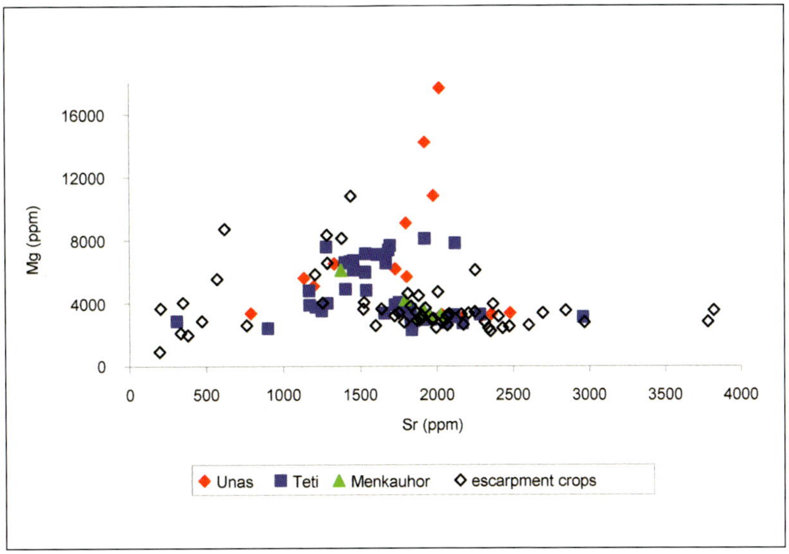

Fig. 159: Mg/Sr-diagram of the pyramids of Unas, Teti, and Menkauhor in comparison with the assumed Saqqara quarry sites. A good coincidence of the plots is evident with the exception of three Unas samples.

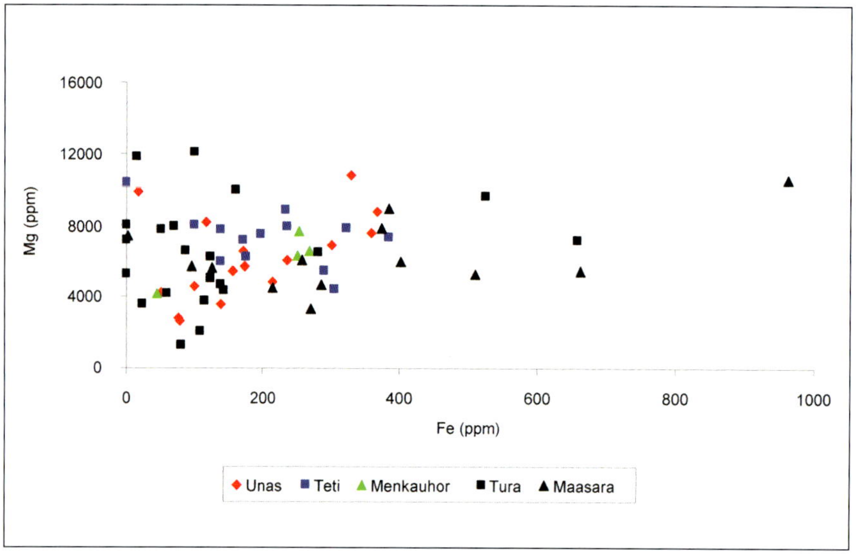

Fig. 160: Mg/Fe-diagram of the pyramids of Unas, Teti, and Menkauhor casing material in comparison with the Tura-Maasara quarry district. A good coincidence with especially Maasara is evident. For Teti a close attribution is less certain. For Menkauhor the sample amount is too low for a certain attribution.

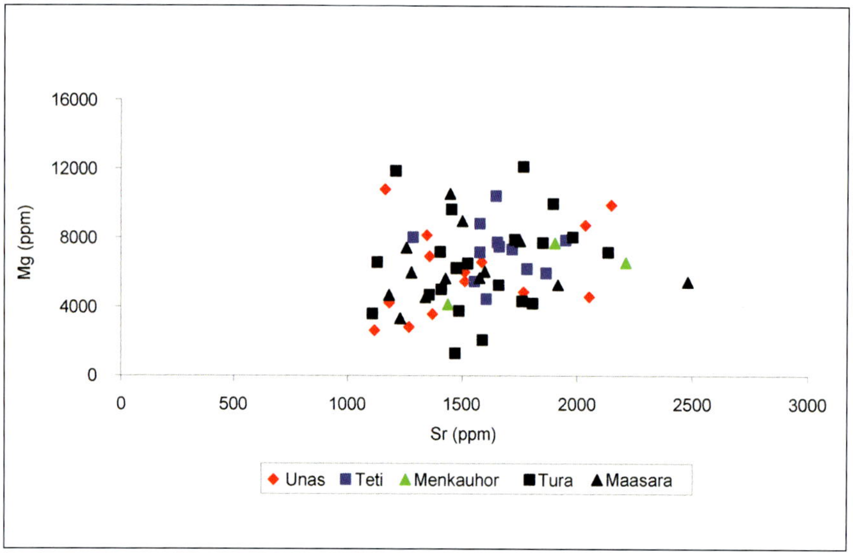

Fig. 161: Mg/Sr-diagram of the pyramids of Unas, Teti, and Menkauhor casing material in comparison with the Tura-Maasara quarry district. Due to the wide overlapping of the Tura-Maasara plots no closer attribution is possible.

Geochemical determination of the casing material

The casing material of the pyramid and limestone blocks of the burial chamber and corridors as well as the fine limestones of the mortuary temple derive, according to Figs. 160–161, from the quarries of Tura-Maasara.

Here, the various casing samples are more similar to the Maasara data, indicating that Unas also received his stones from there as well as from an own quarry, which, according to the geochemical data, cannot be far away from that quarry site used by Snofru.

XIV. The Pyramid of Teti (2347–2337 BC)

Geographical position (entrance): 29°52'31" N, 31°13'18" E

Height: 52.5 m | Base: 78.5 m | Slope: 53°

The pyramid of Teti, the first ruler of the 6th dynasty, is situated in the northern field of the Saqqara necropolis (Fig. 162). LAUER/LECLANT[137] from the Mission Archéologique Française de Saqqara (MAFS) first investigated it, and excavations have continued since then.

The main pyramid has a core of five steps, with subterranean corridors and chambers similar to those of the Djedkare and Unas pyramids. The core masonry building blocks are poorly hewn and with relatively wide joints in between, filled with Nile mud and gravel mortar. The two Saqqara rock types with different deterioration patterns were rather randomly used (Fig. 163).

The entrance of the pyramid was dug into the courtyard of the mortuary temple along the pyramid's north wall. The entrance corridor had a barrier made of three granite blocks, and both the beginning and the end of the corridor was cased with red granite.

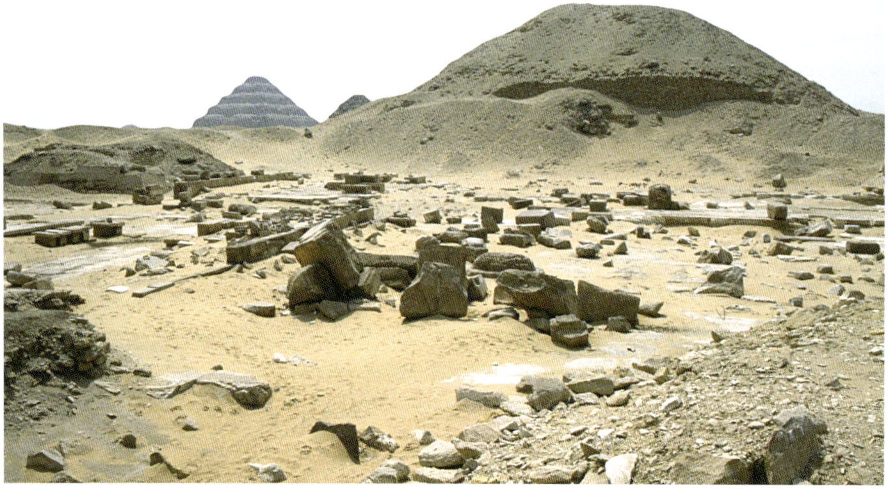

Fig. 162: Teti pyramid with its mortuary temple. Step Pyramid in the background.

Fig. 163: Detail of the core masonry from Teti pyramid. The various only roughly hewn block sizes correspond to the natural layering of the rock. They show different stages of deterioration and they are bound by Nile mud mortar and gravel in wide joints.

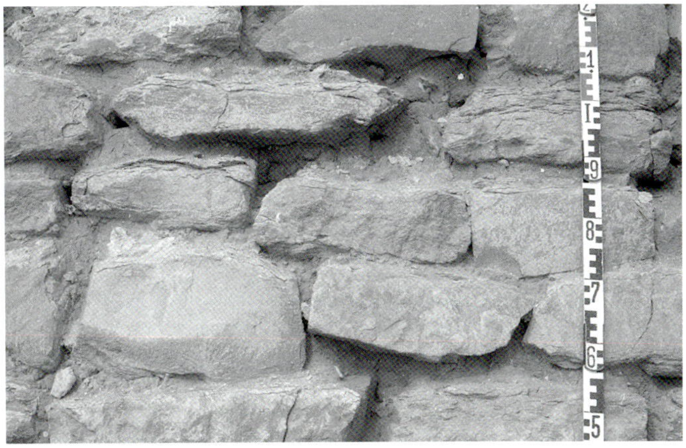

Fig. 164: Calcite-alabaster pavement in Teti's mortuary temple.

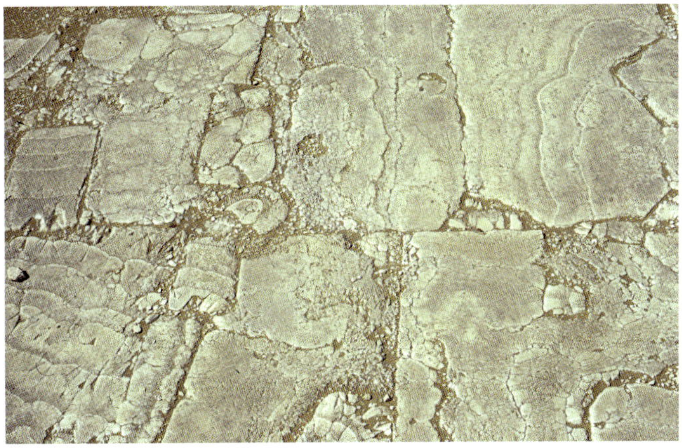

Both of the underground chambers had gabled ceilings made from three layers of huge, fine limestone blocks. The walls of the burial chamber are covered with fine limestone. Both the walls of the burial chamber and antechamber are again inscribed with "pyramid texts".

The valley temple and the 300 m long causeway have not been excavated but are located to the south-east of the pyramid. Part of the causeway is visible where it meets the mortuary temple.

The floor of its entrance hall was paved with thick slabs of polished calcite-alabaster (Fig. 164). In addition, a still remaining staircase and relics of an altar are of the same material. The actual entrance in the middle of the courtyard has a single door over a quartzite doorstep. Quartzite was also used in parts as pavement, doorposts, and thresholds within the many rooms of the temple. In front of the cult pyramid stand two quartzite offering bowls.

A special feature of Teti's pyramid complex is the intensive use of various hard rock types: granite pillars of the colonnaded court are mainly of a light red Aswan quality, and the walls and ceilings are of fine limestone. The false door was made of granite upon a base of quartzite, and the five statue niches were lined with granite blocks. The masses of debris contained a number of basalt blocks up to half a cubic metre in size. The blocks are hard to reconstruct within the temple's architecture.

As is often the case, the main pyramid is surrounded by an open courtyard, which is delimited by an enclosure wall. The cult pyramid at the south-east corner of the main pyramid also has its own enclosure wall.

Outside the main complex are various other ruins from the burials of Teti's family and closest officials. These include the small pyramid complexes of Khuit and Iput I, consorts of Teti.

Petrography

The same bimodality of the stone material, as already described in connection with the Djoser and Userkaf pyramids, can be observed when studying the petrography of the Teti pyramid's core masonry.

The colour of the core masonry stones varies from grey-beige of the more marly types to yellowish-brown of the sandier and fossil-rich types.

With a hand lens, especially the qualities richer in sand contain many fossil remains with a characteristic glassy lustre.

Under microscope, the more marly limestones are extremely fine-grained with grain sizes less than 0.01 mm, consisting partly of extremely fine-milled fossil parts and rarely of intact fossil remains. Darker portions in the matrix are elements rich in bituminous matter and could occasionally be identified as peloids (Fig. 165). The evenly dispersed sand grains are around 0.1 mm in size in the majority of specimens.

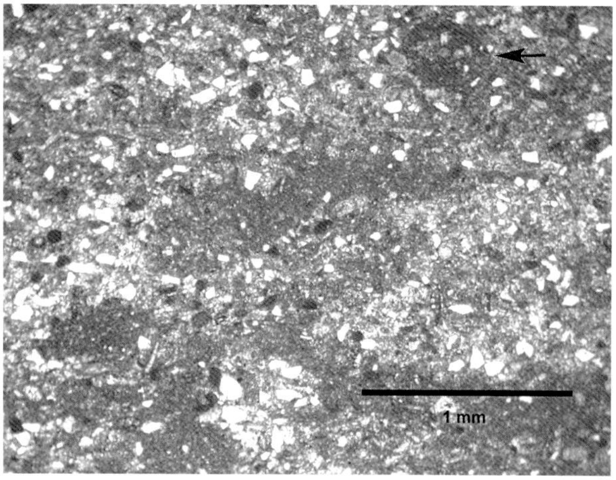

Fig. 165: Micro-photo (T 5) of a marly limestone sample from the core masonry of Teti pyramid. The fine-grained matrix with various darker, bituminous enriched parts and a remaining peloid (black arrow) contains sharp edged sand grains until maximum 0.1 mm (1 pol. filter).

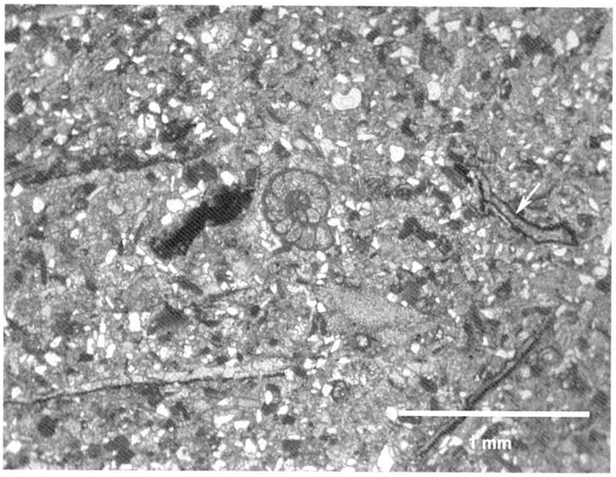

Fig. 166: Micro-photo (T 36) of a sandy limestone sample from the core masonry of Teti pyramid. Within the microsparitic matrix many calcified fossil relicts, partly with a limonitic rim, limonitic dots, and a well-preserved small gastropod in the centre. Some calcified fossil relicts have a black rim of bituminous matter (white arrow) (1 pol. filter).

The more sand-rich limestones normally contain many fossil relics that have decayed within the calcareous matrix. Well-preserved fossils are the exception. Limonite dots give the whole rock a reddish-brown tint. Bituminous matter occurs as black rims around intensively secondary calcified fossil remains (Fig. 166).

Geochemical provenance determination of the core material of the Teti pyramid and subsidiary pyramids

A relatively large number of samples was taken from the pyramid of Teti, expecting a better statistically certified attribution with an increasing number of samples. However, this procedure led only to a denser population within the correlation diagrams, but a more precisely located quarry area hardly could be distinguished.

The geochemical correlation diagrams of Figs. 158–159 clearly show that by far the largest portion of the samples derived from local Saqqara quarry areas with its sequence of sandy and marly limestones of the eastern escarpment, which extends towards the Nile valley. As these various quarry sites cannot be differentiated due to their proximity, further speculations about the precise localization of a certain quarry site will not be made here. In fact, the entire eastern escarpment might be regarded as a potential quarry area, but due to extensive covering of debris, only limited sampling is possible.

From a closer study of the Mg/Fe-diagram in Fig. 158 it can be observed that a number of the plotted data appear outside the Saqqara field in contrast to the Mg/Sr-diagram (Fig. 159), where most of the plotted data appear within this field. This indicates a general relationship of all samples to the Saqqara province. The answer to this discrepancy is similar to the case of the Unas pyramid, where most probably reused stones from the unfinished Sekhemkhet complex were incorporated into the core masonry. These limestone qualities were originally used only as casing material, and their quarry sites had been exhausted since the early years of Snofru's reign.

The building material of the pyramids belonging to the two queens, Khuit and Iput I, known to have lived under Teti, derives, as in the case of the destroyed "unknown pyramid" (Menkauhor ?), exclusively from the quarry area of the eastern escarpment. In addition, the bad state of preservation of these monuments in relation to the pyramids of Unas and Teti might be the result of using less-suited building material.

Geochemical provenance determination of the casing material of the Teti pyramid and subsidiary pyramids

According to the geochemical correlation diagrams of Figs. 160–161, the Tura-Maasara quarry district provided the casing limestone for the pyramids of Teti and his queens. However, a more precise localization is not indicated in these diagrams. Even with a relatively poor number of samples, the casing of the "unknown pyramid" can also be petrographically attributed to the Tura-Maasara area.

XV. The Pyramid of Pepi I (2335–2285 BC) at Saqqara-South

Geographical position (entrance): 29°51'17" N, 31°13'07" E

Height: 52 m | Base: 78 m | Slope: 53°

Pepi I was the second ruler of the 6th dynasty. Today, his pyramid stands in ruins 12 m high (Fig. 167). The site has been excavated by the MAFS for years during the last decade[138].

The pyramid has a core of six steps. It was constructed in the same way as the Djedkare pyramid, i.e. using small blocks of limestone bound with a clay mortar. The pyramid was encased in fine limestone, which today remains intact only at the lowest levels.

The entrance lies in the courtyard pavement next to the pyramid's north face. The subterranean levels are similar to earlier pyramids of the 5th and 6th dynasties, beginning with a descending limestone corridor that leads to the subterranean chambers. The gabled ceilings of the antechamber and burial chamber consist of three layers of huge limestone blocks.

Fragments have been found of a greywacke sarcophagus that once stood on the west wall of the burial chamber.

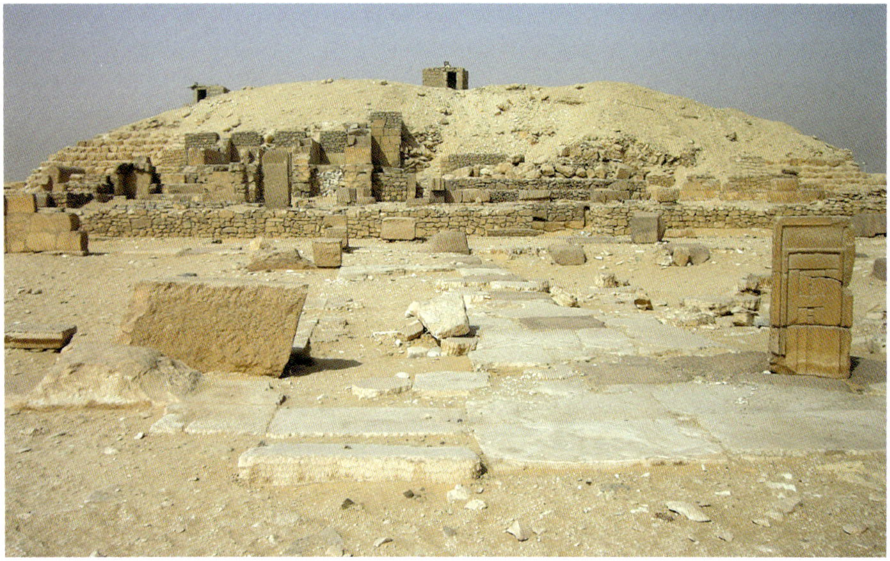

Fig. 167: Pepi I pyramid with remains of the mortuary temple, view from east. Note the large pavement blocks of calcite-alabaster in the foreground.

Pyramid texts not only cover the walls of the antechamber and burial chamber, but also the corridors. Some of these texts remain in place, while other parts are in fragments.

The temple followed the general style of the 6th dynasty as far as the building stones are concerned. The floor of the courtyard is covered with calcite-alabaster, while the pilasters, the statue niches, and the doorways are of Aswan granite.

No information exists on the valley temple and causeway.

At the south-eastern corner of the pyramid complex there is a small cult pyramid, and recently the French mission (MAFS) discovered two small pyramid complexes of Pepi I's consorts on the southern side[139]. The complex has been extensively restored.

Petrography

The bad state of preservation of the pyramid itself is most probably due to intensive stone robbery, whereby a crater-like hole was dug into its central part, mainly to gain material for local lime burning. Consequently, traces of burn-

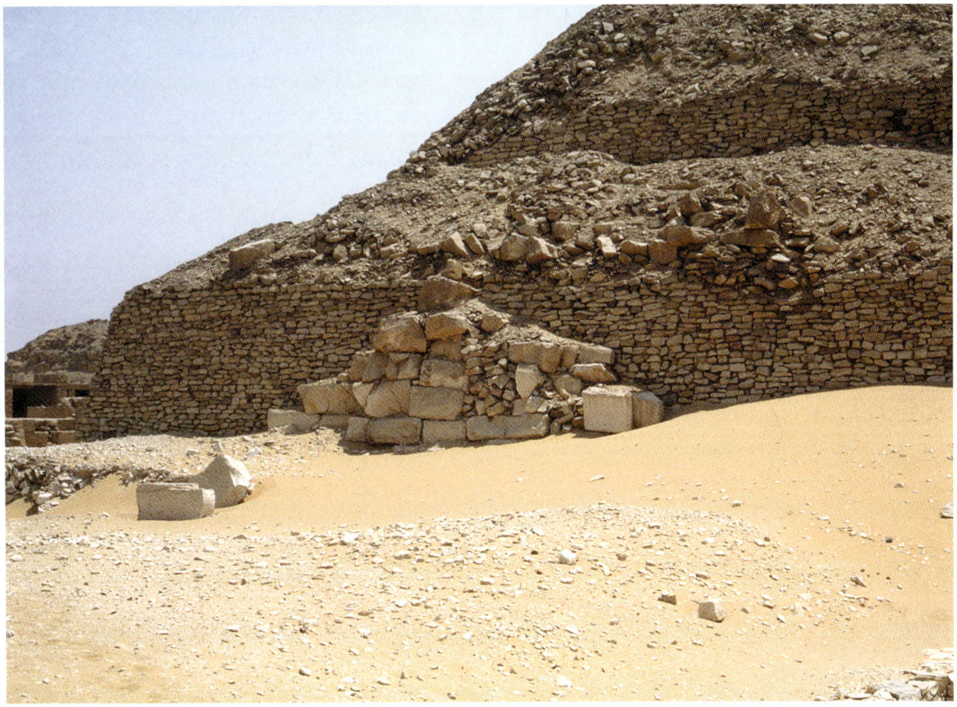

Fig. 168: Small building stones of local origin are characteristic for the core of Pepi I, Pepi II, and Merenre pyramids. The picture shows an example from Pepi II pyramid with remains (restored) of much larger backing stones of non local fine limestone.

ing can be found everywhere, but especially within the area of the former mortuary temple, which was filled up with waste and rubbish when excavation started.

The limestones from the entrance hall, partly heavily burnt and brittle, are in all likelihood from the Maasara quarries (Figs. 154–155).

In the waste dump to the east of the temple excavation, larger blocks of a conglomerate quartzite and some stone slabs of greywacke from Wadi Hammamat could be detected, therefore supporting the assumption that these materials were also used for temple decoration.

This is all the more likely as the sarcophagus in the burial chamber is also of greywacke (a specimen of the sarcophagus was kindly provided by C. BERGER from the MAFS).

Due to the central crater dug into the pyramid, sampling of the core material was unproblematic. Sampling was also facilitated, as the building stones are astonishingly small (Fig. 168). In spite of the geochemical correlation with the Djedkare-Isesi core material, the petrography of the Pepi I core stones differs to some extent. Almost all samples consist of a distinctly fine bedding structure. The sand grains in the more sand-rich varieties are not exclusively of quartz; a small percentage is of plagioclase and muscovite (white mica) and the content of fossil fragments seems to be lower. This leads to the suspicion that Pepi I used his own quarry site, which was situated in close proximity to the other Saqqara-South pyramid owners.

Despite the ruined state of the pyramid and the almost complete dismantling of its casing, sufficient blocks, especially from the destroyed temple area, could be sampled. Already from its macroscopic appearance, this fine limestone may again be attributed to the Tura-Maasara quarry district. The coarser internal structure points towards a source at Maasara. When freshly cut, this limestone gives a light smell of hydrogen sulphide, which is rather uncommon for Maasara limestone and most probably points to a special quarry site within this district.

Geochemical determination of the core material of Pepi I's pyramid

In the correlation diagrams of Figs. 152–153 only a part of the plotted data appears in the field of the assumed quarry areas along the escarpment, and is therefore comparable to the already discussed pyramid of Djedkare-Isesi. This indicates that the actual quarry site of the core masonry of Pepi I's pyramid was not detected. On the other hand, the position of the plotted data supports a general similarity between the pyramid stones and the local outcropping limestones sampled during the course of this investigation. Intensive study of aerial photographs did not result in the detection of a depression in close vicinity of the pyramid which could have been the position of an open pit quarry. However, it is most probably buried under excavation debris today.

Geochemical determination of the casing material of Pepi I's pyramid

The geochemical determination of the fine limestone used for the casing masonry of the Pepi I pyramid is certain: According to the correlation diagrams of Figs. 154–155, the sample values best fit to the quarry area of Maasara. Further, the data also indicate that a specific quarry within the Maasara area was chosen to deliver this special stone material.

XVI. The Pyramid of Merenre (2285–2279 BC)

Geographical position (entrance): 29°51'03" N, 31°12'54" E

Estimated Height: ~ 78 m | Present Height: 52.4 m | Slope: 53.7°

The pyramid, which is mostly in ruins today, is under investigation by the MAFS[140]. It is barely noticeable compared with other nearby monuments. It lies relatively deep in the desert, at the south-western edge of the necropolis. The plan of the pyramid is unclear, and little is known about most of the complex, including the mortuary temple.

The pyramid's entrance is at the north side and leads into a descending corridor. The corridor first meets a vestibule and then continues along a second corridor to the burial chamber, where the decorated greywacke sarcophagus stood on the west wall.

Uni, governor of Upper Egypt, provided valuable information on the origin of the materials used to build the pyramid: Granite from Aswan, calcite-alabaster from Hatnub, and dark greywacke from "Ibhat" (i.e. Wadi Hammamat[141]) were used, the latter for the pyramidion and the sarcophagus[142].

Petrography

Merenre's pyramid is completely covered by rubble and waste and gives at first glance the impression of a huge waste dump. This is most probably the result of intensive stone robbery in the past, where the larger blocks were removed and only the gravel and waste, which fills the interspaces of the building stones, were left. Nevertheless, sufficient samples could be taken to analyse both the provenance of the core and casing material.

Petrographically, the core masonry of Merenre's pyramid is almost identical with that of Djedkare-Isesi's, underlining that it also derived from the hilly plain of Saqqara-South. Possible quarry sites were expected along the northern flank of Wadi Tafla, from which the causeway runs towards a probable port (Fig. 169). However, this is still questionable. Directly south of the pyramid, the aerial photograph shows a blurred square, which might be a sand-filled artificial imbedding, i.e. a possible quarry. This would also explain the geochemical similarities of the pyramid samples with the outcrop samples taken from the Saqqara-South escarpment rock units.

The material cropping out there, at the northern flank of Wadi Tafla, however, contains in addition to limestone banks also layers of a lamellar marly-silty rock, which tends to deteriorate in a foliated pattern and gives the wadi the name "Tafla". Within the pyramid's masonry, this unsuitable building material was rather frequent. This might also be one of the reasons for the poor preservation of Merenre's pyramid and temple constructions, and another argument for the localization of the quarry site in the flank of Wadi Tafla.

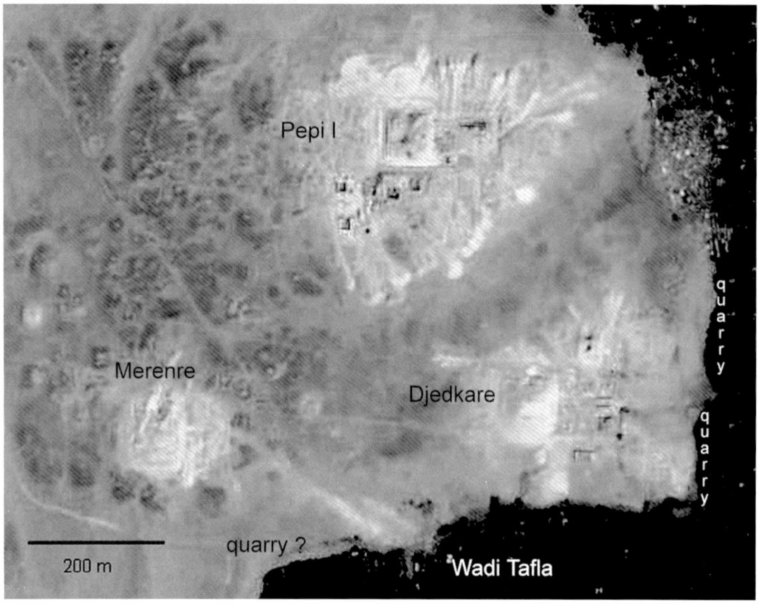

Fig. 169: Overview of Saqqara-South pyramids of Merenre, Pepi I, and Djedkare north of Wadi Tafla (after Google satellite image 2006).

The remaining fine limestone of the casing and the excavated parts of the mortuary temple show macroscopic similarities to Tura-Maasara. On the other hand, the mortuary temple has only been partly excavated, so the use of further stone qualities can only be determined by searching the debris. Alongside the aforementioned limestone types, quartzite and calcite-alabaster fragments were found in the debris, and granite and greywacke were used as architectural elements, i.e. as the architrave of the entrance gate and the pyramidion respectively.

Geochemical determination of the core material

Comparing the plotted geochemical data of the assumed quarry sites of Saqqara-South (Figs. 152–153) with those of the core masonry from Merenre's pyramid, a relatively good correlation is visible, showing that the real quarry site for the core masonry must be in the vicinity.

Geochemical determination of the casing material

The geochemical values of the fine limestone used for the casing of Merenre's pyramid and the inaccessible temple complex fit, according to Figs. 154–155, best to the fields of the Maasara quarries. This result corresponds well with the casing stone type used by Djedkare-Isesi and Pepi I.

With a more representative geochemical sampling of the various gallery quarries in Maasara, specific quarry sites used by each king could certainly be identified, but unfortunately, this district is strictly off limits to non-military persons.

XVII. The Pyramid of Pepi II (2279–2219 BC)

Geographical position (entrance): 29°50'26" N, 31°12'48" E

Height: 52.5 m | Base: 79 m | Slope: 53°

The pyramid of Pepi II (Fig. 170) at Saqqara-South was the last to be built in the tradition of the Old Kingdom. The pyramid is located on the southern edge of the necropolis, south of Wadi Tafla and close to Mastabat el-Faraoun. It was systematically investigated by G. Jequier between 1936 and 1940.

The pyramid is built like those of Pepi's predecessors using small pieces of local limestone bound with a clay and gravel mortar for the core, and fine limestone for the casing. The core consisted of five steps. After the casing was completed and the north chapel had been built, a ca. 7 m wide band of bricks was added around the pyramid at the level of the third layer of core blocks, possibly in order to strengthen the lower levels of the casing.

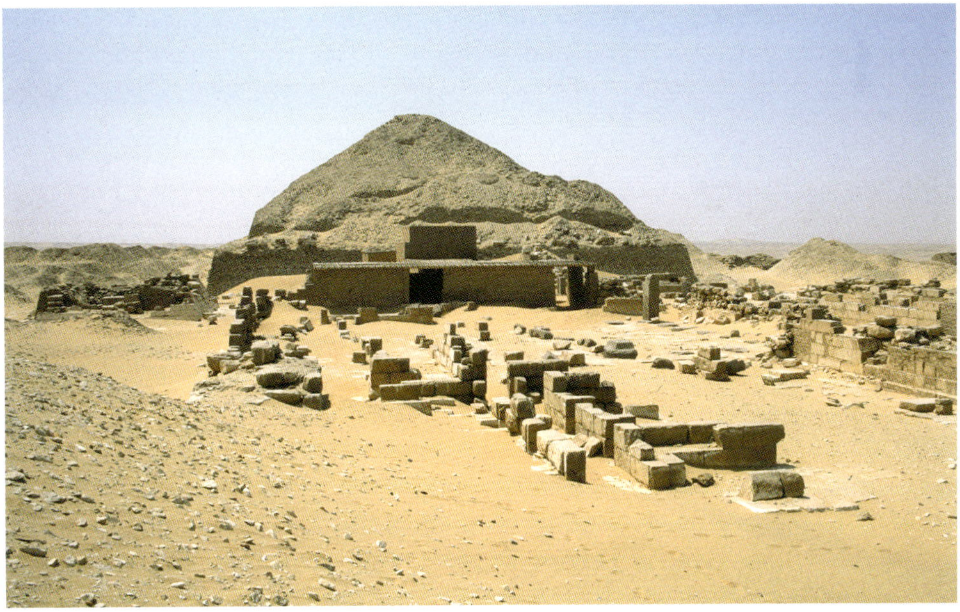

Fig. 170: Pepi II pyramid with its temple seen from east.

The subterranean chambers of the pyramid are again decorated with "pyramid texts", this time in the corridor leading to the inner chambers. A huge sarcophagus of Aswan granodiorite is still in the burial chamber. Some fragments of calcite-alabaster and diorite vases were found.

The valley temple of Pepi II is different from those of earlier kings. It is a wide structure, with an angular terrace open on its east side, which was most probably connected to a harbour[143].

The approx. 400 m long causeway connects the valley temple with the mortuary temple, located east of the pyramid. The mortuary temple is not unlike those of earlier 5th and 6th dynasty rulers. The entrance corridor leads to an open courtyard that was paved with limestone and adorned with eighteen pillars of conglomerate quartzite. One of these has survived, but had to be extensively restored.

At the south-eastern corner of the pyramid's courtyard stood a small cult pyramid.

To the north, north-west, and south of Pepi II's complex, the pyramids of three of his queens were found: Neith, Iput II, and Udjebten.

Petrography

The petrography of the core masonry stones differs to some extent from the three pyramids north of Wadi Tafla (Fig. 169).

Macroscopically, the stones represent the typical Saqqara sequence with an irregular alternate layering of more massive, sand-rich limestones and finer, less sandy and marly limestones.

Under microscope, the sand-rich types can be separated into two varieties. The first variety is the well-known sandy limestone, rich in fossil fragments, which are partly calcified and fringed by bituminous matter. The second is a coarser-grained, calcareous sandstone, dominated by quartz grains of 0.5–0.05 mm in size but also contains hornblende, epidote, plagioclase, and microcline (Fig. 172). This relatively immature sandstone is new in the Saqqara petrographic record and requires a more detailed provenance attribution

Geochemical provenance determination of the core material

The stone material of the core masonry of Pepi II's pyramid as well as of the subsidiary pyramids is of a rather poor sandy limestone quality as far as the consistency and deterioration nature are concerned. The bimodal character of this building material has already been mentioned in the macroscopic description. Its poor resistance to weathering was already acknowledged during the construction period, as a kind of supporting masonry of fine, durable limestone of the same quality as the casing limestone described below was incorporated.

The geochemical values of the presumable local building material are presented in Figs. 152–153 and are compared with the data from the assumed quarry areas in Saqqara-South. Similar to the three pyramids north of Wadi Tafla it becomes apparent that a certain relationship but not a satisfying correspondence of the core material with these quarry areas exists. Again, this demonstrates that local material was used for the pyramid constructions but that the exact localization of the individual quarry that provided the core masonry could not be determined.

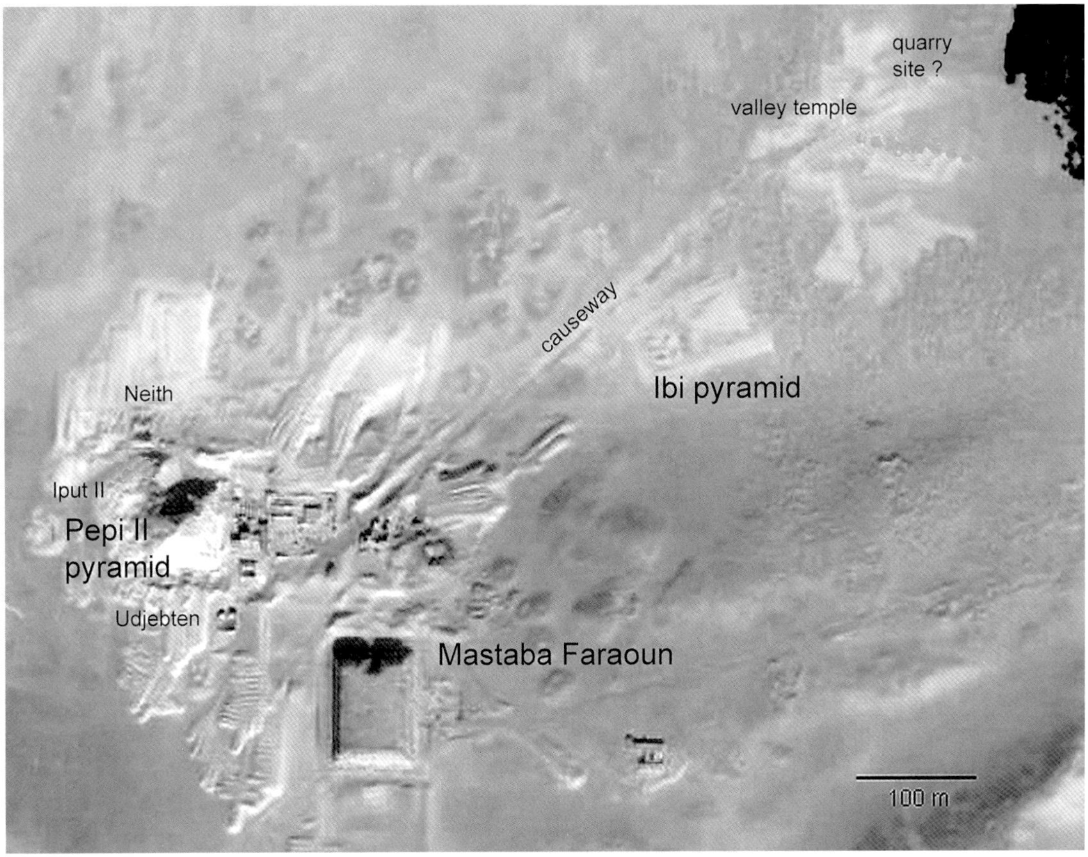

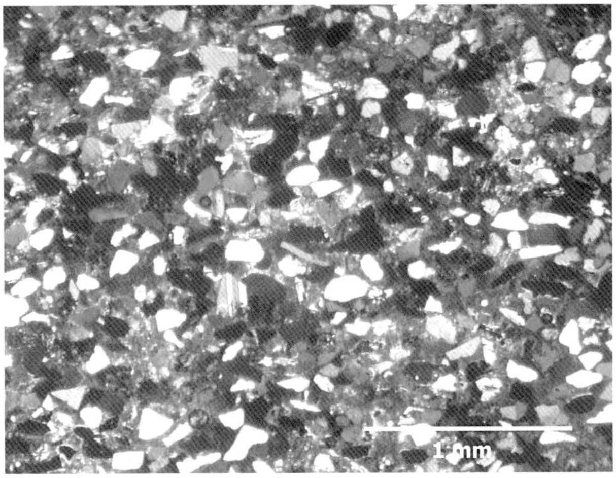

▲ *Fig. 171: Modified satellite image (Google 2006) showing Pepi II- and Mastabat el-Faraoun monuments in Saqqara-South. The valley temple with harbour (?) is cut into the flat rock slope, a possible quarry position?*

Fig. 172: Micro-photo (P II 41) of calcareous sandstone of Pepi II core masonry. The sand grains are mainly quartz, but partly also plagioclase, both embedded in a micritic-calcitic matrix (crossed polarizing filters).

Unfortunately, no quarry site containing these rock varieties could be detected in the pyramid's vicinity. A possible quarry site could be the harbour-like area in front of the valley temple, but it is doubtful whether the necessary quantities of stone could have been gained from there. Neither a stereoscopic aerial photograph nor the satellite image gives any indication of another nearby quarry area (Fig. 171). Another explanation could be that the original quarry is, again, covered by the excavation debris, extending north of the pyramid.

Geochemical provenance determination of the casing material

The fine limestones were not only used as casing and backing stones for the pyramids themselves, but also for the various temples and to support the inner chambers.

The provenance determination is thus of special interest. The geochemical correlation diagrams of Figs. 154–155 show, equivalent with those of Djedkare Isesi, Pepi I and Merenre in Saqqara-South, close similarities to the quarry district of Maasara. Again, the distribution of the plotted data points to an individual quarry of Pepi II, the last king of the Old Kingdom, who built a stone pyramid.

XVIII. Concluding Remarks

Summarizing the petrographical and geochemical information provided by this study on the provenance of the building materials of the Old Kingdom pyramids, the stone derived in most cases, at least where the core material is concerned, from the close vicinity of the pyramids. The few exceptions are as follows:

The Meidum pyramid is located at a distance of about 1 km to its quarry site. In this case, the elevated position of the pyramid in relation to the low topographical level of the quarry site may have been one of the reasons for the selection of the pyramid's position.

The Mastabat el-Faraoun needed an approx. 2.5 km long drag ramp to its quarry area. This, of course, is unusual, even when the selected building material was easy to extract. Further, due to the relatively low weight of this stone material, the transport of even large block sizes along a more or less horizontal plain could be achieved without any major difficulties.

Finally, the long distance of the two sun temples in Abu Gurob to their quarry sites, i.e. of about 1 km (Userkaf) and 1.6 km (Niuserre) is also quite far. In these cases, the rather long distance to the Saqqara-North and Abusir quarry sites, from where they gained the building material, was insignificant as the amount was relatively low.

For all the other pyramids and funerary complexes, the close proximity between quarry site and monument is striking. A pyramid's position was determined by the availability and closeness of sufficient rock material, irrespective of political criteria, like the distance to the capital. From a political point of view, it is at first glance somewhat puzzling why Snofru selected the rather remote site of Dahshur for his two pyramids. However, sufficient building material was available in the surrounding area of the monuments, and in the case of the Red Pyramid, even relatively easy to obtain.

The knowledge of the solid limestone rock unit of the Gizeh plateau was undoubtedly the reason for the construction of three great pyramids after the failure of the Bent Pyramid. This site was then exclusively used for these colossal monuments and the tombs of family members and retinue. Lastly, the almost inexhaustible sources of high quality building material may have been a major argument in favour of this site.

On the other hand, the masses of considerably sized stone that had to be transported from the remote granite sources south of Aswan have to be taken into account. Further, the extraction and transport of the many other rock types used in the building process such as calcite-alabaster, greywacke, quartzite, basalt, and anorthosite from mainly remote sites of the eastern and western deserts requires immense logistical organization, which seemingly caused no major difficulties for the ancient Egyptians.

In the case of the 3rd dynasty pyramids not only the core, but also the casing material came from nearby local sources. However, from Snofru on, the casing and backing limestones were always extracted from the eastern escarpment, where most probably each king opened his own individual and exclusive gallery. The transport distance between 25–35 km and the quarry works required immense logistical effort. Especially at Tura and Maasara (and in the case of Khufu also at Mokattam hill), the engineers selected only the best weather-resistant stone qualities. The mining operations were done with such excellent skill that most of the quarries remain stable until today.

Notes

1 "backing stones" stand for one or two stone layers between the core masonry and the outer casing stones, different of the core material and mostly of the same provenance as the casing material.
2 EVERS 1929.
3 KLEMM/KLEMM 1993.
4 KLEMM/KLEMM 1991.
5 E.g. LAUER 1985.
6 WILDUNG 1977.
7 WILDUNG 1969.
8 LAUER 1936–1965.
9 STADELMANN 1997.
10 KLEMM/KLEMM/MURR 1998.
11 BLANCKENHORN 1921.
12 EGSMA 1983.
13 KLITZSCH/LIST/PÖHLMANN 1987.
14 YOUSSEF ET AL. 1984.
15 YOUSSEF ET AL. 1984, Abb. 4.
16 EGSMA 1983.
17 KLITZSCH/LIST/PÖHLMANN 1987.
18 YOUSSEF ET AL. 1984.
19 E.g. LAUER 1936; STADELMANN 1997.
20 LAUER 1962.
21 E.g. STADELMANN 1997.
22 KLEMM/KLEMM 1993.
23 GONEIM 1957.
24 WISSA 1997.
25 GONEIM 1957.
26 REISNER/FISHER 1911.
27 REISNER/FISHER 1911.
28 LEHNER 1996.
29 BARSANTI 1901.
30 LEHNER 1996.
31 BARSANTI 1906; BARSANTI 1907; BARSANTI 1912.
32 VON BECKERATH 1997.
33 YOUSSEF ET AL. 1984.
34 DREYER/KAISER 1980.
35 E.g. RICKE 1950; YOYOTTE 1963; LAUER 1967; WILDUNG 1969; FAKHRY 1969; PARRA ORTIZ 1996.
36 EL-KHOULI 1991.
37 STADELMANN 1987.
38 STADELMANN 1980; STADELMANN 1997.
39 PETRIE 1892; PETRIE 1910.
40 BORCHARDT 1928.
41 DREYER/KAISER 1980.
42 FAKHRY 1959.
43 STADELMANN 1982; STADELMANN1997.
44 KLEMM/KLEMM 1993.
45 STADELMANN 1997.
46 KLITZSCH/LIST/PÖHLMANN 1987.
47 HERMINA/KLITZSCH/LIST 1989.
48 EGSMA 1983.
49 EGSMA 1983.
50 YOUSSEF ET AL. 1984.
51 STADELMANN 1982; STADELMANN 1997.
52 VON ZITTEL 1883.
53 BLANKENHORN 1921.
54 CUVILLIER 1934.
55 HUME 1925.
56 WAGNER 1935.
57 KNETSCH 1953.
58 SAID 1962.
59 SAID 1971.
60 STROUGO 1985.
61 YEHIA 1985.
62 AIGNER 1982; AIGNER 1983.
63 GAURI 1984.
64 LEHNER 1985a; LEHNER 1985b; LEHNER 1986.
65 LEHNER 1985c.
66 AIGNER 1982.
67 GAURI 1984.
68 GAURI 1984.
69 AIGNER 1982; AIGNER 1983.
70 STROUGO 1979.
71 YEHIA 1985.
72 AIGNER 1982.
73 YEHIA 1985.
74 YEHIA 1985.
75 LEHNER 1985a; LEHNER 1985b.
76 AIGNER 1982.
77 YEHIA 1985.
78 HAWASS 1998; HAWASS 2005.
79 LEHNER 1985a; LEHNER 1985b.
80 REISNER 1942.
81 LEHNER/ALLEN/GAURI 1980; AIGNER 1982.
82 REISNER 1942.
83 REISNER 1942.
84 HAWASS 1998.
85 AIGNER 1982.
86 DAVIDOVITS/MORRIS 1988.

[87] CAMPBELL/FOLK 1991; CAMPBELL/FOLK 1992.
[88] HARRELL/PENROD 1993.
[89] MORRIS 1992; MORRIS 1993; MORRIS 1994; MORRIS 2004.
[90] BARSOUM/GANGULY/HUG 2006.
[91] DAVIDOVITS/MORRIS 1988.
[92] ARNOLD 1991.
[93] KLEMM/KLEMM 1998.
[94] HAWASS 1997; HAWASS 1998; HAWASS 2005.
[95] GREENOUGH/GORTON/MALLORY-GREENOUGH 2001.
[96] KLEMM/KLEMM 1993; HARRELL 1995.
[97] HOFFMEIER 1993.
[98] MOORES 1991.
[99] GOEDICKE 1972.
[100] SALAH EL-NAGGAR 2005.
[101] MARAGIOGLIO/RINALDI 1963–1975.
[102] REISNER 1942.
[103] AIGNER 1983.
[104] KLEMM/KLEMM 1993.
[105] KLEMM/KLEMM 1993; KLEMM/KLEMM 2008.
[106] YEHIA 1985.
[107] REISNER 1931.
[108] REISNER 1942.
[109] KLEMM/KLEMM 1981.
[110] MONTET 1946.
[111] MARAGIOGLIO/RINALDI 1962; MARAGIOGLIO/RINALDI 1963–1975.
[112] VALLOGIA 2004.
[113] KLEMM/KLEMM/MURR 1998.
[114] VALLOGIA 1997.
[115] EGSMA 1983.
[116] KLEMM/KLEMM/MURR 1998.
[117] VALLOGIA 2004.
[118] STADELMANN 1997.
[119] JEQUIER 1928.
[120] JEQUIER 1928; GOMAÀ 1973.
[121] SWELIM 1983.
[122] KLEMM/KLEMM/MURR 1998.
[123] LAUER 1955.
[124] BORCHARDT 1905.
[125] VON BISSING/KEES 1922–1928.
[126] RICKE 1965–1969.
[127] BORCHARDT 1910; BORCHARDT 1913.
[128] BORCHARDT 1909.
[129] VERNER 1995.
[130] VERNER 1995.
[131] VERNER 1997.
[132] VERNER 1985.
[133] BORCHARDT 1907.
[134] VERNER 2005.
[135] MOURSI 1987; MOURSI 1988a; MOURSI 1988b.
[136] YOUSSEF ET AL. 1984.
[137] LAUER/LECLANT 1972.
[138] LABROUSSE 1996.
[139] LABROUSSE/ALBOUY 1999.
[140] LAUER 1972–1973; LAUER 1983.
[141] KLEMM/KLEMM 1993.
[142] WISSA 1994.
[143] KLEMM/KLEMM/MURR 1998.

References

The following and more recent comprehensive studies on Egyptian pyramids are recommended for further and detailed information:

Z. HAWASS, *The pyramids of ancient Egypt, Carnegie Museum of Natural History, The Carnegie Series on Egypt*, Pittsburgh 1990.

P. JANOSI, *Die Pyramiden: Mythos und Archäologie*, München 2004.

M. LEHNER, *The Complete Pyramids*, London 1997.

W. M. F. PETRIE, *The pyramids and temples of Gizeh. New and revised edition* (ed. by Z. HAWASS), London 1990.

R. STADELMANN, *Die großen Pyramiden von Giza*, Graz 1990.

R. STADELMANN, *Die Ägyptischen Pyramiden – Vom Ziegelbau zum Weltwunder*, Mainz 1997.

M. VERNER, *The Pyramids: The Mystery, Culture, and Science of Egypt's Great Monuments*, London 1997.

AIGNER 1982
T. AIGNER, *Zur Geologie und Geoarchäologie des Pyramidenplateaus von Gizeh, Ägypten, Natur und Museum* 112, Frankfurt/M. 1982, pp. 377–388.

AIGNER 1983
T. AIGNER, *Facies and origin of nummulitic builtups; an example of the Giza Pyramids plateau (Middle Eocene, Egypt)*, in: *Neues Jahrbuch für Geologie Paläontologie-Abhandlungen* 166, 1983, pp. 347–368.

ARNOLD 1991
D. ARNOLD, *Building in Egypt: Pharaonic Stone Masonry*, New York 1991, p. 316.

BARSANTI 1901
A. BARSANTI, *Ouverture de la pyramide de Zaouiét el-Aryan*, in: *ASAE* 2, 1901, pp. 92–94.

BARSANTI 1906
A. BARSANTI, *Fouilles de Zaouiét el-Aryan*, in: *ASAE* 7, 1906, pp. 257–286.

BARSANTI 1907
A. BARSANTI, *Fouilles de Zaouiét el-Aryan*, in: *ASAE* 8, 1907, pp. 201–210.

BARSANTI 1912
A. BARSANTI, *Fouilles de Zaouiét el-Aryan*, in: *ASAE* 12, 1912, pp. 57–63.

BARSOUM/GANGULY/HUG 2006
M. W. BARSOUM/A. GANGULY/G. HUG, *Microstructural Evidence of Reconstituted Limestone Blocks in the Great Pyramids in Egypt*, in: *Journal of the American Ceramic Society* 89/12, 2006, pp. 3788–3796.

VON BECKERATH 1971
J. VON BECKERATH, *Abriss der Geschichte des Alten Ägypten*, Oldenburg/München/Wien 1971.

VON BECKERATH 1997
J. VON BECKERATH, *Chronologie des Pharaonischen Ägypten*, Mainz 1997.

VON BISSING/KEES 1922
F. W. VON BISSING/H. KEES, *Untersuchungen zu den Reliefs aus dem Re-Heiligtum des Rathures* I, München 1922.

VON BISSING/KEES 1923
F. W. VON BISSING/H. KEES, *Das Re-Heiligtum des Königs Ne-Woser-Re (Rathures)* II: *Die kleine Festdarstellung*, Leipzig 1923.

VON BISSING/KEES 1928
F. W. VON BISSING/H. KEES, *Das Re-Heiligtum des Königs Ne-Woser-Re (Rathures)* III: *Die große Festdarstellung*, Leipzig 1928.

BLANCKENHORN 1921
M. P. BLANCKENHORN, *Handbuch der Regionalen Geologie* VII, Abt. 9/23: *Ägypten*, Heidelberg 1921.

BORCHARDT 1905
L. BORCHARDT, in: F. W. VON BISSING, *Das Re-Heiligtum des Königs Ne-Woser-Re (Rathures)* I: *Der Bau*, Leipzig 1905.

BORCHARDT 1907
L. BORCHARDT, *Das Grabdenkmal des Königs Ne-user-re*, *WVDOG* 7, Leipzig 1907.

BORCHARDT 1909
L. BORCHARDT, *Das Grabdenkmal des Königs Nefer-ir-ke-re*, *WVDOG* 11, Leipzig 1909.

BORCHARDT 1910–1913
L. BORCHARDT, *Das Grabdenkmal des Königs Sahure*, *WVDOG* 14, Leipzig 1910–1913.

BORCHARDT 1928
L. BORCHARDT, *Die Entstehung der Pyramide an der Baugeschichte der Pyramide bei Meidum nachgewiesen*, Beiträge *Bf* 1, Cairo 1928.

CAMPBELL/FOLK 1991
D. H. CAMPBELL/R. L. FOLK, *The Ancient Egyptian Pyramids – Concrete or rock?*, in: *Concrete International* 13/8, 1991, pp. 28–39.

CAMPBELL/FOLK 1992
D. H. CAMPBELL/R. L. FOLK, *Are the Pyramids of Egypt Built of Poured Concrete Blocks?*, in: *Journal of Geological Education* 40, 1992, pp. 25–34.

CUVILLIER 1934
J. CUVILLIER, *Un niveau très fossilifère à la base de l'éocene supérieur, au sud de la pyramide de Menkara*, in: *Bulletin de Société Géologique de France* 5, 4/1–3, 1934, pp. 61–67.

DAVIDOVITS/MORRIS 1988
J. DAVIDOVITS/M. MORRIS, *The Pyramids – An Enigma Solved*, New York 1988.

DREYER/KAISER 1980
G. DREYER/W. KAISER, *Zu den kleinen Stufenpyramiden Ober- und Mittelägyptens*, in: *MDAIK* 36, 1980, pp. 43–59.

EGSMA 1983
EGYPTIAN GEOLOGICAL SURVEY AND MINING AUTHORITY (ed.), *The Geological Map of Greater Cairo*, Cairo 1983.

EL-KHOULI 1991
A. EL-KHOULI, *Meidum, Australian Centre for Egyptology, Reports* 3, Sydney 1991.

EVERS 1929
H. G. EVERS, *Staat aus dem Stein*, München 1929.

FAKHRY 1959a
A. FAKHRY, *The monuments of Snofru at Dahshur* I: *The Bent Pyramid*, Cairo 1959.

FAKHRY 1959b
A. FAKHRY, *The monuments of Snofru at Dahshur* II: *The Red Pyramid*, Cairo 1959.

FAKHRY 1969
A. FAKHRY, *The Pyramids*, Chicago 1969.

GAURI 1984
K. L. GAURI, *Geologic study of the Sphinx*, in: *NARCE* 127, 1984, pp. 24–43.

GOEDICKE 1972
H. GORDICKE, *Re-used Blocks from the Pyramid of Amenemhat I at Lisht, PMMA, Egyptian Expedition* 20, New York 1972.

GOMAÀ 1973
F. GOMAÀ, *Chaemwese - Sohn Ramses' II. und Hoherpriester von Memphis*, *ÄA* 27, Wiesbaden 1973.

GONEIM 1957
Z. GONEIM, *Horus Sekhem-khet. The Unfinished Step Pyramid at Saqqara* I, Cairo 1957.

GREENOUGH/GORTON/MALLORY-GREENOUGH 2001
J. D. GREENOUGH/M. P. GORTON/L. M. MALLORY-GREENOUGH, *The Major- and Trace Element Whole-Rock Fingerprints of Egyptian Basalts and the provenance of Egyptian Artefacts*, in: *Geoarchaeology* 16/7, 2001, pp. 763–784.

HARRELL/PENROD 1993
J. A. HARRELL/B. E. PENROD, *The great Pyramid debate – Evidence from the Lauer sample*, in: *Journal of Geological Education* 41, 1993, pp. 358–369.

HARRELL 1994
J. A. HARRELL, *Harrell's response to Morris Article*, in: *Journal of Geological Education* 42, 1994, pp. 195–198.

HARRELL/BOWN 1995
J. A. HARELL/T. M. BROWN, *An Old Kingdom Basalt Quarry at Widan el-Faras and the Quarry Road to Lake Moeris*, in: *JARCE* 32, 1995, pp. 71–91.

HAWASS 1997
Z. HAWASS, *The Discovery of the Harbors of Khufu and Khafra at Giza*, in: C. BERGER-EL-NAGGAR/B. MATHIEU (éds.), *Ètudes sur l'Ancien Empire et la nécropole de Saqqara dédieés à Jean-Philippe Lauer, Orientalia Monspeliensia* 9, Montpellier 1997, pp. 245–256.

HAWASS 1998
Z. HAWASS, *Pyramid Construction. New Evidence Discovered at Giza*, in: H. GUKSCH/D. POLZ (Hrsg.), *Stationen. Beiträge zur Kulturgeschichte Ägyptens* (FS Stadelmann), Mainz 1998, pp. 53–62.

HAWASS 2005
Z. HAWASS, *Khufus National Project. The Great Pyramid of Giza in the year 2528 BC*, in: P. JANOSI (ed.), *Structure and Significance. Thoughts on Ancient Egyptian Architecture, Untersuchungen der Zweigstelle Kairo des Österreichischen Archäologischen Institutes* 25, *Österreichische Akademie d. Wissenschaften, Denkschriften* 33, Wien 2005, pp. 305–330.

HERMINA/KLITZSCH/LIST 1989
M. HERMINA/E. KLITZSCH/F. LIST, *Stratigraphic Lexicon and Explanatory Notes to the Geological Map of Egypt 1:500 000*, Cairo 1989.

HOFFMEIER 1993
J. K. HOFFMEIER, *The use of Basalt in Floors of Old Kingdom Pyramid Temples*, in: *JARCE* 30, 1993, pp. 117–123.

HUME 1911
W. F. HUME, *The effect of secular oscillation in Egypt during the Cretacious and Eocene periods*, in: *(Quarterly) Journal of the Geological Society* 67, 1911, pp. 118–148.

HUME 1925
W. F. HUME, *Geology of Egypt* I: *The surface features of Egypt, their determing causes and relation to geological structure*, Cairo 1925.

JEQUIER 1928
G. JEQUIER, *Le Mastabat Faraoun*, *Fouilles à Saqqarah* 9, Le Caire 1928.

JEQUIER 1936–1940
G JEQUIER, *Le Monument Funeraire de Pepy II*, *Fouilles à Saqqarah* 13, Le Caire 1936–1940.

KLEMM/KLEMM 1981
R. KLEMM/D. KLEMM, *Die Steine der Pharaonen*, München 1981.

KLEMM/KLEMM 1991
R. KLEMM/D. KLEMM, *Mortar Evolution in the Old Kingdom of Egypt*, in: *Archaeometry* 90, 1991, pp. 445–454.

KLEMM/KLEMM 1993
R. KLEMM/D. KLEMM, *Steine und Steinbrüche im Alten Ägypten*, Heidelberg 1993.

KLEMM/KLEMM 1998
R. KLEMM/D. KLEMM, *Die Integralrampe als Konstruktionselement großer Pyramiden*, in: H. GUKSCH/D. POLZ (Hrsg.), *Stationen. Beiträge zur Kulturgeschichte Ägyptens* (FS Stadelmann), Mainz 1998, pp. 87–95.

KLEMM/KLEMM/MURR 1998
R. KLEMM/D. KLEMM/A. MURR, *Zur Lage und Funktion von Hafenanlagen an den Pyramiden des Alten Reiches*, in: *SAK* 26, 1998, pp. 173–189.

KLEMM/KLEMM 2008
R. KLEMM/D. KLEMM, *Stones and Quarries in Ancient Egypt*, London 2008.

KLITZSCH/LIST/PÖHLMANN 1987
E. KLITZSCH/F. K. LIST/G. PÖHLMANN, *Geological Map of Egypt, NH 36 SW Beni Suef*, Cairo 1987.

KNETSCH 1953
G. KNETSCH, *Zur Geologie eines altägyptischen Baudenkmals in Giza*, in: *Neues Jahrbuch für Geologie und Paläontologie-Abhandlungen* 3, 1953, pp. 112–115.

LABROUSSE/LAUER/LECLANT 1978
A. LABROUSSE/J.-P. LAUER/J. LECLANT, *Le temple haut du complexe funéraire du roi Ounas*, *BdE* 73, Le Caire 1978.

LABROUSSE/MOUSSA 1996
A. LABROUSSE/A. MOUSSA, *Le temple d'accueil du complexe funéraire du Roi Ounas*, *BdE* 111, Le Caire 1996.

LABROUSSE 1996
A. LABROUSSE, *The pyramids of Pepi and his queens at Saqqara*, in: *EA* 8, 1996, pp. 3–6.

LABROUSSE/ALBOUY 1999
A. LABROUSSE/A. ALBOUY, *Les Pyramides des Reines: Une Nouvelle Nécropole à Saqqara*, Paris 1999.

LAUER 1936–1965
J.-P. LAUER, *La pyramide à degrés*, *Fouilles à Saqqara* 14, Le Caire 1936–1965.

LAUER 1955
J.-P. LAUER, *Le temple haut de la pyramid de Ouserkaf à Saqqara*, in: *ASAE* 53, 1955, pp. 119–133.

LAUER 1962
J.-P. LAUER, *Histoire monumentale des pyramides d'Égypte* I: *Les pyramides à degrés (IIIe dynastie)*, *BdE* 39, Le Caire 1962.

LAUER 1967
J.-P. LAUER, *Sur la pyramide de Meidoum et les deux pyramides du roi Snefrou à Dahchour*, in: *Orientalia* 36, 1967, pp. 239–254.

LAUER 1972–1973
J.-P. LAUER, *Nouvelles Recherches à la Pyramide de Meren-Re*, in: *BIE* 53–54, 1972–1973, pp. 63–69.

LAUER 1985
J.-P. LAUER, *A propos de l'invention de la pierre de taille par Inhotep pour la demeure d'éternité du roi Djoser*, in: P. POSENER-KRIÉGER (éd.), *Mélanges Gamal Eddin Mokhtar*, *BdE* 97, Le Caire 1985, pp. 61–67.

LAUER/LECLANT 1972
J.-P. LAUER/J. LECLANT, *Le temple haut du complexe funéraire du roi Teti*, *BdE* 51, Le Caire 1972.

LECLANT/CLERC 1988
J. LECLANT/G. CLERC, *Fouilles et travaux en Égypte et au Soudan 1986–1987*, in: *Orientalia* 57, 1988.

LEHNER/ALLEN/GAURI 1980
M. LEHNER/J. P. ALLEN/K. L. GAURI, *The ARCE Sphinx Project: A Preliminary Report*, in: *NARCE* 112, 1980, pp. 3–33.

LEHNER 1985a
M. LEHNER, *The Giza Plateau Mapping Project: Season 1984–85*, in: *NARCE* 131, 1985, pp. 23–56.

LEHNER 1985b
M. LEHNER, *A contextual approach to the Giza pyramids*, in: *AfO* 32, 1985, pp. 136–158.

LEHNER 1985c
M. LEHNER, *The Development of the Giza Necropolis: The Khufu Project*, in: *MDAIK* 41, 1985, pp. 109–143.

LEHNER 1986
M. LEHNER, *The Giza Plateau Mapping Project: Season 1986*, in: *NARCE* 135, 1986, pp. 29–55.

LEHNER 1996
M. LEHNER, *Z 500 and the Layer Pyramid of Zawiyet el-Aryan*, in: P. DER MANUELIAN (ed.), *Studies in honour of William Kelly Simpson* II, Boston 1996, pp. 507–522.

MARAGIOLIO/RINALDI 1962
V. MARAGIOGLIO/C. A. RINALDI, *Notizie sulle piramidi di Zedefra, Zedkara Isesi, Teti*, Torino 1962.

MARAGIOLIO/RINALDI 1963–1975
V. MARAGIOGLIO/C. A. RINALDI, *L'Architettura delle Piramidi Menfite, testo & tavole*, Torino-Rapallo 1963–1975.

MOORES 1991
R. G. MOORES, *Evidence for use of a stone-cutting drag saw by the fourth dynasty Egyptians*, in: *JARCE* 28, 1991, pp. 139–148.

MOURSI 1987
M. I. MOURSI, *Die Ausgrabungen in der Gegend um die Pyramide des Djed-Ka-Ra „Issj" bei Saqqara*, in: *ASAE* 71, 1987, pp. 185–193.

MOURSI 1988a
M. I. MOURSI, *Die Ausgrabungen in der Gegend um die Pyramide des Djed-Ka-Ra „Issj" bei Saqqara* II, in: *GM* 105, 1988, pp. 65–68.

MOURSI 1988b
M. I. MOURSI, *Die Ausgrabungen in der Gegend um die Pyramide des Djed-Ka-Ra „Issj" bei Saqqara* III, in: *GM* 106, 1988, pp. 65–72.

MOUSTAFA/YEHIA/ABDEL TAWAB 1985
A. R. MOUSTAFA/M. A. YEHIA/S. ABDEL TAWAB, *Structural setting of the areas east of Cairo, Maadi, and Helwan*, in: *Sci. Res. Ser.* 5, 1985, pp. 40–64.

MONTET/JOUBÉ-LAGÜE 1946
P. MONTET/G. JOUBÉ-LAGÜE, *Abou Roach*, Paris 1946.

MORRIS 1992
M. MORRIS, *Geopolymeric Pyramids, a Rebuttal to R. L. Folk and D. H. Campbell*, in: *Journal for Geological Education* 40, 1992, pp. 35–46.

MORRIS 1993
M. MORRIS, *How not to Analyse Pyramid Stones – The Invalid Conclusion of James A. Harrell and Bret E. Penrod*, in: *Journal of Geological Education* 41, 1993, pp. 364–369.

MORRIS 1994
M. MORRIS, *Response to J. A. Harrell*, in: *Journal of Geological Education* 42, 1994, pp. 198–203.

MORRIS 2004
M. MORRIS, *The Egyptian Pyramid Mystery Is Solved*, Detroit 2004.

PARRA ORTIZ 1996
J. M. PARRA ORTIZ, *Houni et Snéfrou: Les pyramides de Meidoum et Dahchour*, in: *GM* 154, 1996, pp. 77–91.

PETRIE 1892
W. M. F. PETRIE, *Medum*, London 1892.

PETRIE/MACKAY/WAINWRIGHT 1910
W. M. F. PETRIE/E. MACKAY/G. WAINWRIGHT, *Meydum and Memphis* III, *BSAE* 16, London 1910.

REISNER/FISHER 1911
G. A. REISNER/C. FISCHER, *The Harvard University-Museum of Fine Arts Egyptian Expedition*, *BMFA* 9, no. 50, Boston 1911, pp. 13–20.

REISNER 1931
G. A. REISNER, *Mycerinus. The Temples of the Third Pyramid at Giza*, Cambridge 1931.

REISNER 1942
G. A. REISNER, *A History of the Giza Necropolis* I, Cambridge 1942.

RICKE 1950
H. RICKE, *Bemerkungen zur Baukunst des Alten Reiches* II, *Beiträge Bf* 5, Cairo 1950, pp. 1–128.

RICKE 1965–69
H. RICKE (Hrsg.), *Das Sonnenheiligtum des Königs Userkaf* I: *Der Bau*, *Beiträge Bf* 7, Cairo 1965.

H. RICKE (Hrsg.), *Das Sonnenheiligtum des Königs Userkaf* II: *Die Funde*, *Beiträge Bf* 8, Wiesbaden 1969.

SAID 1962
R. SAID, *The Geology of Egypt*, Amsterdam 1962.

SAID 1971
R. SAID, *Explanatory notes to accompany the geological map of Egypt*, *Geological Survey of Egypt*, Paper No. 56, Cairo 1971.

EL-NAGGAR 2005
S. EL-NAGGAR, *Les couvrements de la granite dans les Pyramides de Giza*, in: P. JANOSI (ed.), *Structure and Significance. Thoughts on Ancient Egyptian Architecture*, *Untersuchungen der Zweigstelle Kairo des Österreichischen Archäologischen Institutes* 25, *Österreichische Akademie der Wissenschaften, Denkschriften* 33, Wien 2005, pp. 425–438.

STADELMANN 1980
R. STADELMANN, *Snofru und die Pyramiden von Meidum und Dahschur*, in: *MDAIK* 36, 1980, pp. 437–449.

STADELMANN 1982
R. STADELMANN, *Die Pyramiden des Snofru in Dahschur. Erster Bericht über die Ausgrabungen an der nördlichen Steinpyramide*, in: *MDAIK* 38, 1982, pp. 109–123.

STADELMANN 1987
R. STADELMANN, *Beiträge zur Geschichte des Alten Reiches. Die Länge der Regierung des Snofru*, in: *MDAIK* 43, 1987, pp. 229–240.

STADELMANN 1990
R. STADELMANN, *Die großen Pyramiden von Giza*, Graz 1990.

STADELMANN 1985
R. STADELMANN, *Die ägyptischen Pyramiden: Vom Ziegelbau zum Weltwunder*, *Kulturgeschichte der antiken Welt* 30, Mainz 1985.

STROUGO 1979
A. STROUGO, *The Middle Eocene-Upper Eocene boundary in Egypt*, in: *Annals of the Geological Survey of Egypt* 9, 1979, pp. 454–469.

STROUGO 1985
A. STROUGO, *Eocene stratigraphy of the eastern Greater Cairo/Gebel Mokattam-area*, in: *Sci. Res. Ser.* 5, 1985, pp. 1–39.

SWELIM 1983
N. M. SWELIM, *Some problems on the History of the Third Dynasty*, *Archaeological and Historical Studies* 7, Alexandria 1983.

VALLOGGIA 1997
M. VALLOGGIA, *La descenderie de la pyramide de Radjedef á Abu Rawash*, in: C. BERGER-EL-NAGGAR/B. MATHIEU (éds.), *Études sur l'Ancien Empire et la necropole de Saqqara dédieés à Jean-Philippe Lauer*, *Orientalia Monspeliensia* 9, Montpellier 1997, pp. 417–425.

VALLOGGIA 2004
M. VALLOGGIA, *Le complexe funéraire de Rêdjedef à Abou Rawash: Etat des travaux après dix campagnes (1995–2004)*, in: *BSFE* 161, 2004, pp. 12–27.

VERNER 1985
M. VERNER, *Les sculptures de Rêneferef découvertes á Abousir*, in: *BIFAO* 85, 1985, pp. 267–280.

VERNER 1995
M. VERNER, *The Pyramid Complex of Khentkaus*, *Abusir* 3, Prag 1995.

VERNER 1997
M. VERNER, *The Pyramids: The Mystery, Culture, and Science of Egypt's Great Monuments*, London 1997.

VERNER 2005
M. VERNER, *Abusir pyramids quarry and supply road*, in: P. JANOSI (ed.), *Structure and significance. Thoughts on Ancient Egyptian*

architecture, *Untersuchungen der Zweigstelle Kairo des Österreichischen Archäologischen Institutes* 25, *Österreichische Akademie der Wissenschaften, Denkschriften* 33, Wien 2005, pp. 531–538.

WAGNER 1935
G. WAGNER, *Geologische Beobachtungen an den Pyramiden*, in: *Aus der Heimat. Naturwissenschaftliche Monatsschrift* 48/1, 1935, pp. 19–21.

WILDUNG 1969
D. WILDUNG, *Die Rolle ägyptischer Könige im Bewusstsein ihrer Nachwelt* I, *MÄS* 17, München 1969.

WILDUNG 1977
D. WILDUNG, *Imhotep und Amenhotep – Gottwerdung im alten Ägypten*, *MÄS* 36, München 1977.

WISSA 1994
M. WISSA, *Le sarcophage de Merenrê et l'éxpédition à Ibhat* (I), in: C. BERGER EL-NAGGAR (éd.), *Études pharaoniques. Hommages à Jean Leclant* I, *BdE* 106, Le Caire 1994, pp. 379–387.

WISSA 1997
M. WISSA, *A propos du sarcophage de Sékhemkhet*, in: C. BERGER EL-NAGGAR/B. MATHIEU (éds.), *Études sur l'Ancien Empire et la nécropole de Saqqâra dédiées à Jean-Philippe Lauer* II, *Orientalia Monspeliensia* 9, Montpellier 1997, pp. 445–448.

YEHIA 1985
M. A. YEHIA, *Geological Structures of the Gizeh Pyramid plateau*, in: *Sci. Res. Ser.* 5, 1985, pp. 100–120.

YOUSSEF ET AL. 1984
M. YOUSSEF/O. CHERIF/M. BOUKHARY/A. MOHAMED, *Geological Studies on the Saqqara Area, Egypt*, in: *Neues Jahrbuch für Geologie und Paläontologie-Abhandlungen* 168/1, 1984, pp. 125–144.

VON ZITTEL 1883
K. A. VON ZITTEL (Hrsg.), *Beiträge zur Geologie und Paläontologie der libyschen Wüste und der angrenzenden Gebiete von Ägypten*, *Palaeontographica* 30, Kassel 1883.

For short descriptions of the various pyramids, we also refer to the following internet sites:

http://oi.uchicago.edu/OI/PROJ/GIZ/Giza.html.
http://www.touregypt.net/featurestories/htm.

List of abbreviations

ÄA = Ägyptologische Abhandlungen, Wiesbaden
AfO = Archiv für Orientforschung, Wien
ASAE = Annales du Service des Antiquités de L'Égypte, Le Caire
BdE = Bibliothèque d'Étude, Institut Français d'Archéologie Orientale, Le Caire
Beiträge Bf = Beiträge zur Ägyptischen Bauforschung und Altertumskunde, Cairo/Wiesbaden
BIE = Bulletin de l'Institut d'Égypte, Le Caire
BIFAO = Bulletin de l'Institut Français d'Archéologie Orientale, Le Caire
BSAE = British School of Archaeology in Egypt, London
BSFE = Bulletin de la Société Française d'Égyptologie, Paris
EA = Bulletin of the Egypt Exploration Society, London.
GM = Göttinger Miszellen. Beiträge zur ägyptologischen Diskussion, Göttingen
IFAO = Institut Français d'Archéologie Orientale du Caire, Le Caire
JARCE = Journal of the American Research Centre in Egypt, Boston
MÄS = Münchener Ägyptologische Studien, Berlin/München
MDAIK = Mitteilungen des Deutschen Archäologischen Instituts, Abteilung Kairo, Mainz/Berlin
NARCE = Newsletter of the American Research Centre in Egypt, Cairo
SAK = Studien zur Altägyptischen Kultur, Hamburg